LOST CREATURES
OF THE EARTH

CONTENTS

TABLES

ACKNOWLEDGMENTS

The author thanks the following organizations for providing photographs for this book: the National Aeronautics and Space Administration (NASA), the National Museums of Canada, the National Optical Astronomy Observatories (NOAO), the National Park Service, the U.S. Geological Survey (USGS), the U.S. Navy, and the Woods Hole Oceanographic Institution. The author also thanks Frank K. Darmstadt, Senior Editor, and the rest of the editorial and production staff for their invaluable contribution to the book.

FOREWORD

One of the most fascinating aspects of geology is the ability to view ancient "monsters" that once roamed the Earth. These monsters can be viewed only as fossil imprints in rocks or fossilized bones and shells. However, artists' renditions and movies bring them to life. The realization that life was very different at one time immediately raises the question of why these creatures disappeared. Indeed, at times in Earth history, 75 percent or more of the species on the planet catastrophically died off. The great mass extinction of the dinosaurs is now popularly believed to have been caused by an asteroid colliding with the Earth. The dinosaur mass extinction, however, is not the most profound extinction, which is not associated with any extraterrestrial influence. What else can kill off so many species at one time? *Lost Creatures of the Earth* by Jon Erickson investigates just that question to shed some light on this baffling and frightening phenomenon. Should we be worried about perishing in a mass extinction?

In order to characterize the ages and complexity of life, chapter 1 contains a brief synopsis of historical geology. This chapter is important because it places the dominant life-forms into historical perspective. It also shows the amazing variety of life. Chapters 2 and 3 describe the various animals, marine and terrestrial, and their various habitats and habits. Chapter 4 recounts the major extinction events through Earth history. It lists most of the major species that went extinct in each event and possible geologic and climatic mechanisms for the extinctions. Chapter 5 investigates the major proposed mechanisms for extinction events both in their development and the effect on different forms of life. Chapters 6 and 7 explore the aftermath of these catastrophic events. How do some animals survive the events, and how do they evolve to utilize the ecologic niches abandoned by the extinct species? These chapters are relatively biologic. Chapter 8 describes the various cycles, both terrestrial and extraterrestrial, around which life must adapt. Much of nature is controlled by these cycles. Finally, chapters 9 and 10 describe the oddities of the fossil record as well as the living record. These include odd-looking animals, animals with strange abilities and habits, and animals in remote habitats.

The book is written in a friendly style with more description than complex science. Descriptions of individual species and habitats are fascinating and of wide interest. The book is a novel approach to an interesting and fascinating topic. As with all of Erickson's books, it contains a useful glossary for those who do not understand all of the terms. A reference list is also included for those who wish to investigate the topic further.

—Alexander E. Gates, Ph.D.

INTRODUCTION

Perhaps no other force has influenced the development of life on Earth as much as the mass extinction of species. Extinction has weeded out the weak in favor of the strong, ensuring that life continues to survive even after the worst calamities nature can devise. During its long history, the Earth has endured global environmental disasters that have eliminated more than half the species living at the time. All extinctions indicate biologic systems in extreme stress caused by radical changes in the environment, possibly caused by huge asteroid or comet impacts or by powerful volcanic eruptions.

Every mass extinction marked a watershed in the evolution of life, when whole groups of organisms disappeared and were replaced by entirely new species. Consequently, extinction plays a pivotal role in the development of life. It is an inevitable part of evolution and is especially important for the progress of life-forms. Therefore, the extinction of species has been almost as prevalent as their origination. Had species not become extinct to make room for advanced organisms, life on Earth would not have progressed to the myriad of biologic forms now in existence.

Many evolutionary pathways weave along branches of the tree of life. Species leave their footprints in the fossil record, itself only a modest representation of all those that have gone before. Practically every conceivable biologic form and function have lived, some more successful than others. Through this trial-and-error method of speciation, natural selection has selected certain species to prosper while failures were left on the trash heap of extinction.

This book follows the historical development of life over geologic time and examines plants and animals that inhabit the seas and occupy the land. It follows life's progress from the sea onto the land and its evolution in a terrestrial environment. After examining life-forms in all their seemingly limitless varieties, the text concentrates on mass extinctions in Earth history and places the extinction of species into their historical context. It then examines some possible factors that influenced the mass disappearance of species.

The discussion continues with an examination of how mass extinctions have influenced the outcome of life on Earth and the processes that affected the evolutionary development of species. It then looks at the various factors that have influenced the growth of life on this planet since the beginning and how these have shaped living organisms down through geologic history. Next, it takes a journey back in time in search of the strangest creatures that have ever lived on Earth. Finally, it searches for life in some of the most severe living conditions our planet has to offer.

Students of geology and Earth science will find this a valuable reference book to further their studies. Readers will enjoy this clear and easily readable text that is well illustrated with compelling photographs, clearly drawn illustrations, and helpful tables. A comprehensive glossary is provided to define difficult terms, and a bibliography lists references for further reading. Science enthusiasts will particularly enjoy this fascinating subject and gain a better understanding of how the forces of nature have shaped our living Earth.

LOST CREATURES
OF THE EARTH

1

HISTORICAL GEOLOGY
THE AGES OF LIFE

This chapter follows the historical development of life over geologic time. The history of the Earth is divided into segments of geologic time, often defined by mass extinctions of species. The Precambrian era, spanning the first 4 billion years, or roughly 90 percent of geologic history (Table 1), began with only simple creatures living in the sea. By the end of the era, however, an explosion of new, highly specialized species set the stage for more modern life-forms to follow. The Precambrian is divided into the Azoic eon from about 4.6 to 4 billion years ago, the Archean eon from about 4 to 2.5 billion years ago, and the Proterozoic eon from about 2.5 to 0.6 billion years ago.

The last half-billion years, called the Phanerozoic eon, are divided into three eras. They marked major junctures in geologic time and signified the extinction and the evolution of species. The Paleozoic era, from about 570 million to 250 million years ago, witnessed the origin of practically all major marine phyla (organisms that share similar body forms) and the first land dwellers. The Mesozoic era, from about 250 million to 65 million years ago, saw the ascension of the dinosaurs, which held dominion over the world for

TABLE 1   THE GEOLOGIC TIME SCALE

Era	Period	Epoch	Age (millions of years)	First Life-forms	Geology
Cenozoic	Quaternary	Holocene	0.01		
		Pleistocene	2	Humans	Ice age
		Pliocene	11	Mastodons	Cascades
	Tertiary	Neogene			
		Miocene	26	Saber-tooth tigers	Alps
		Oligocene	37		
		Paleogene			
		Eocene	54	Whales	
		Paleocene	65	Horses, alligators	Rockies
Mesozoic	Cretaceous		135		
	Jurassic		190	Birds	Sierra Nevada
				Mammals	Atlantic
				Dinosaurs	
	Triassic		250		
Paleozoic	Permian		280	Reptiles	Appalachians
	Pennsylvanian		310	Trees	Ice age
	Carboniferous				
	Mississippian		345	Amphibians, insects	Pangaea
	Devonian		400	Sharks	
	Silurian		435	Land plants	Laurasia
	Ordovician		500	Fish	
	Cambrian		570	Sea plants, Shelled animals	Gondwana
			700	Invertebrates	
Proterozoic			2,500	Metazoans	
			3,500	Earliest life	
Archean			4,000		Oldest rocks
			4,600		Meteorites

some 170 million years. The Cenozoic era, from 65 million years ago to the present, marked the sudden rise of the mammals, culminating with the arrival of our own species.

THE AGE OF BACTERIA

When surface conditions stabilized on the formative Earth, life began just as soon as it possibly could. Primitive life forms living in the early Precambrian mostly consisted of bacteria, single-celled algae, and algal communities that built layered structures called stromatolites (Fig. 1), from the Greek *stroma* meaning "stony carpet." The stromatolite colonies formed from layers of cells topped by photosynthetic organisms that multiplied using sunlight and supplied the lower layers with nutrients.

For hundreds of millions of years since life began, the only organisms were prokaryotes, from the Greek *karyon* meaning "nutshell." The prokaryotic

Figure 1 Stromatolite of the Missoula group at Glacier National Park, Montana.

(Photo by R. Rezak, courtesy USGS)

Figure 2 *Tube worms on the East Pacific Rise.*

(Courtesy Woods Hole Oceanographic Institution)

cell lacked a distinct nucleus, lived in oxygen-poor conditions, and depended mainly on outside sources for nutrients. The earliest organisms were probably sulfur-metabolizing bacteria similar to those presently living in the tissues of tube worms inhabiting the deep ocean floor near sulfurous hydrothermal vents (Fig. 2). Sulfur would have been abundant on the early, hot planet, spewing from a profusion of volcanoes mostly on the bottom of the ocean. Such an environment well protected from meteorites and strong cosmic radiation could have spurred the evolution of life perhaps as early as 4.2 billion years ago.

Hot surface temperatures on the early Earth caused sulfur atoms to combine into ring molecules in the primordial atmosphere, which blocked out solar ultraviolet radiation. Otherwise, the first living cells would have met with a sizzling death as they basked in the deadly rays of the Sun. However, an ultraviolet shield might not have been necessary since some primitive living forms of bacteria appear to tolerate high levels of ultraviolet radiation.

Bacteria represent life's greatest success story. They occupy a wider domain of environments and span a broader range than any other group of organisms. They are extremely adaptable, indestructible, astoundingly diverse, and absolutely necessary for the existence of other life forms. The bacterial mode of life has been stable from the very beginning of the fossil record. It will no doubt continue as long as the Earth endures, even though most other species will have perished.

During the first half of the Precambrian, the atmosphere and ocean lacked significant levels of oxygen, which was less than 1 percent of the air molecules comprised mostly of nitrogen and carbon dioxide. Therefore, in order to survive under these oxygen-free conditions, bacteria had to obtain energy by the chemical reduction of abundant sulfate ions. The growth of primitive bacteria was thus limited by the amount of nutrients in the soup of organic molecules constantly being generated in the sea around them. Although this form of energy was satisfactory for the time being, bacteria were neglecting a potential energy source—namely sunlight.

The first photosynthesizing plants were proalgae, intermediates between bacteria and blue-green algae. As early as 1 billion years after the Earth's formation, microorganisms called cyanobacteria began using sunlight as their primary energy source. The cells exploited the energy of sunlight to extract from water molecules the hydrogen they needed for self-construction, leaving oxygen as a by-product. The development of photosynthesis was possibly the single most important step in the evolution of life because it provided primitive blue-green algae with a virtually unlimited source of energy.

Photosynthesis also dramatically increased the oxygen content of the ocean and atmosphere (Fig. 3 and Table 2). The oxygen content jumped significantly between 2.2 and 2 billion years ago. During that time, the ocean's high concentration of dissolved iron, which depleted free oxygen to form iron oxide, similar to the rusting of steel, was deposited onto the seafloor and created the world's great iron ore reserves. Around this time, the Earth experienced its first major period of glaciation. The cold ocean waters caused the iron to settle out of suspension.

Well-preserved multicellular algae and a variety of fossil spores discovered in late Precambrian and Cambrian sediments suggest complex sea plants had evolved by this time. However, no other significant remains exist. Fossils of early plants appear to be almost entirely composed of algae much like present-day stromatolites and algal mats on the seashore. Some marine algae lived in the intertidal zone and were exposed out of water for only brief periods to avoid dehydration in the simmering Sun.

THE AGE OF WORMS

The first animals were unicellular, with a committee of simpler cellular organs about the size of prokaryotic cells incorporated into the cell in a symbiotic (living together) relationship. Indeed, symbiosis played a critical role in the evolution of complex organisms. This advancement led to a new type of

Figure 3 *Evolution of life and oxygen in the atmosphere.*

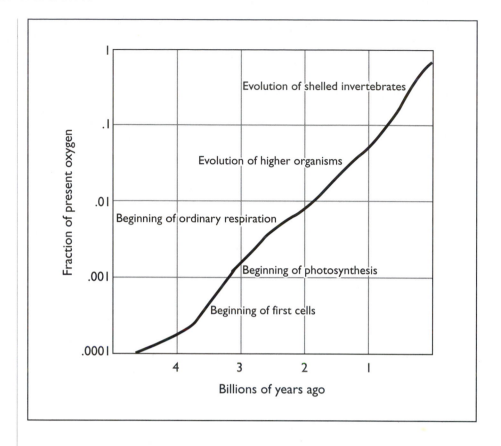

TABLE 2   EVOLUTION OF LIFE AND THE ATMOSPHERE

Evolution	Origin (millions of years)	Atmosphere
Humans	2	Nitrogen, oxygen
Mammals	200	Nitrogen, oxygen
Land animals	350	Nitrogen, oxygen
Land plants	400	Nitrogen, oxygen
Metazoans	700	Nitrogen, oxygen
Sexual reproduction	1,100	Nitrogen, oxygen, carbon dioxide
Eukaryotic cells	1,400	Nitrogen, carbon dioxide, oxygen
Photosynthesis	2,300	Nitrogen, carbon dioxide, oxygen
Origin of life	3,800	Nitrogen, methane, carbon dioxide
Origin of Earth	4,600	Hydrogen, helium

organism called a eukaryote, which evolved from the prokaryotes as early as 3 billion years ago. The eukaryotic cells were typically some 10,000 times larger in volume than prokaryotes. They contained a nucleus that systematically organized genetic material, substantially increasing the number of mutations and the rate of evolution.

Protists were early, single-celled animals that shared many characteristics with plants at a time when no clear demarcations existed between the plant and animal kingdoms. The ability to move about under their own power essentially separates animals from plants, although some animal species perform this function only in the larval stage and become sedentary or fixed to the seabed as adults. Mobility enabled animals to feed on plants and other animals, thereby establishing a new predator-prey relationship.

The development of large metazoans (multicellular organisms) began with the evolution of the early worms. Before about 700 million years ago, the fossil record lacked track-making animals. Afterward, the worms became prolific bottom dwellers and left a preponderance of fossilized tracks, trails, and burrows (Fig. 4). The evolution of primitive segmented worms resulted from the development of muscles and other rudimentary organs, including sense organs and a central nervous system to process the information. The coelomic or hollow-bodied worms burrowed into the seafloor sediments and apparently evolved into higher forms of animal life.

A rounded worm from 565 million years ago was symmetrical from side to side and had what appeared to be a head. It contained three tissue layers, an ectoderm, mesoderm, and endoderm, wrapped around a gut cavity. The body apparently housed a primitive heart, a blood and vascular system, and a

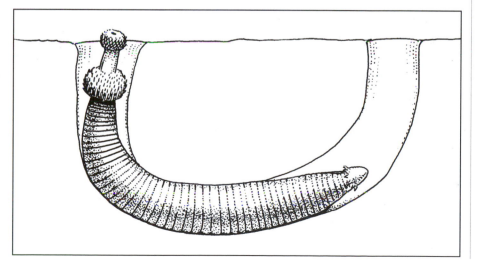

Figure 4 *The early worms left tracks, trails, and burrows.*

means of respiration. The worm left behind fossil burrows, fecal pellets, and scratches in sediments made by the very beginnings of limbs. It possibly gave rise to animals with mouths known as deuterostomes, including chordates, sea urchins, starfish, and marine worms. This ancestral creature might also have led to the protostomes, including mollusks, insects, spiders, leeches, and earthworms.

Some marine worms grew into giants nearly 3 feet long but less than a tenth of an inch thick, thereby providing a large surface area upon which to absorb oxygen and nutrients directly from seawater. The unusually flattened bodies of many animals might have been in response to a limited food supply. In addition, a high surface area to volume ratio collected more sunlight for photogenic algae, which lived symbiotically within their host's body.

Tiny marine worms called acoels might be the closest living representatives of the first bilaterally symmetrical organisms on Earth. They are flatworms grouped along with parasites such as tapeworms and liver flukes. They represent a living relic of the transition between radially symmetrical animals such as jellyfish and more complex bilateral organisms such as vertebrates and arthropods. The animal appears to be a bilateral survivor from before the so-called "Cambrian explosion," a period of unprecedented growth providing a living window onto early metazoan life. Acoels probably branched off from an ancestral animal after the radial jellyfish but before the three major bilateral groups encompassing vertebrates, mollusks, and arthropods began to diverge. Therefore, these offer a living link between primitive and more complex animals.

The rapid evolution of species at the end of the Precambrian was triggered by ecologic stress, geographic isolation resulting from continental drift, and climatic change. Sea levels rose and flooded large portions of the land. The extended shoreline spurred the explosion of new species, producing twice as many phyla living during the Cambrian than before or since. Never were so many experimental organisms in existence, none of which have any modern counterparts. Most new species of the early Cambrian (Fig. 5) were short-lived, however, and soon became extinct.

Organisms no longer relied entirely on surface absorption of oxygen. Gills and circulatory systems evolved when oxygen levels reached about 10 percent of the present value. Life forms rapidly evolved. Unique and bizarre creatures roamed the ocean depths. This great diversity of species followed a late Precambrian ice age, the worst glaciation on Earth—nearly half the land surface was covered by glaciers. When the ice retreated and the seas began to warm, life took off in all directions.

Their fossil impressions are etched in rocks of the Ediacara Formation in South Australia, which were deposited just before the end of the Precambrian. The soft-bodied organisms were mostly marine worms, unusual naked arthropodlike animals, and early relatives of corals and jellyfish. Few, however, looked

Figure 5 *Early Cambrian marine fauna.*

anything remotely like modern animals. The proliferation of these new organisms, representing nearly every major group of marine species, set the stage for the evolution of the progenitors of all animal life on Earth today.

Many unique and bizarre creatures occupied the ocean. The highest percentage of experimental organisms, animals that evolved unusual characteristics, came into being at this time than during any other interval of Earth history. As many as 100 phyla existed, whereas only about a third as many phyla are living today. This biologic exuberance set the stage for more advanced species. Also, for the first time, fossilized remains of animals became abundant due to the evolution of lime-secreting organisms that constructed hard shells.

THE AGE OF CRUSTACEANS

The early Paleozoic was a period of exceptional evolutionary prosperity. It culminated with the first appearance of complex animals clad in exoskeletons within the first 5 to 10 million years. The period witnessed the decline of soft-bodied faunas, which were unsuitable for preservation in the fossil record, and the proliferation of shelly forms, which were much better candidates for fossilization. The new developmental scheme enabled microscopic animals to grow beyond a few thousand cells. Never before or since were so many experimental and unusual species in existence, few of which have any representatives in today's living world.

This high degree of biologic proliferation peaked about 540 million years ago as the ocean filled with a rich assortment of marine life. Seemingly out of nowhere and in bewildering abundance, animals arose in a surprisingly short time cloaked in an astonishing array of body armor. The sudden profusion of protective shells was in response to the newfound ability of animals to prey on each other.

This period followed the worst ice age the world has ever endured, when ice sheets covered half the landmass. It was also a time when oxygen concentrations in the ocean rose to significant levels, spurring the development of large animals. This, in turn, markedly increased the rate of evolution. When the ice retreated and the seas warmed again, life proliferated on an unprecedented scale. Some of the strangest creatures that ever lived dominated the sea, providing a greater number of experimental organisms than during any other interval of geologic history.

As seawater concentrations of calcium increased, early soft-bodied creatures developed a wide variety of skeletons that became more diverse and elaborate over time. Skeletons also evolved in response to an incoming wave of fierce predators, most of which ironically were soft bodied and therefore did not fossilize well. The evolution of skeletons dramatically raised the number of organisms preserved in the fossil record. All known animal phyla that readily fossilized appeared in the lower Cambrian, after which the number of new classes sharply dropped.

The introduction of hard skeletal parts marked the greatest discontinuity in Earths history. It signaled a major evolutionary change by accelerating the developmental pace of new organisms. Nearly all major groups of modern animals appeared in the fossil record. For the first time, animals were equipped with shells, skeletons, legs, and sensing organs. Furthermore, a stable environment enabled marine life to flourish and disperse to all parts of the world.

So many new and varied life forms evolved on such a rapid scale that the period is referred to as the Cambrian explosion. Prior to this spurt of species diversification, the first multicellular animals appeared 600 million years ago in an ocean still dominated by single-celled plankton and bacteria. The seascape abruptly changed with the sudden arrival of animals possessing hard skeletal parts.

Most phyla of living organisms appeared almost simultaneously, many of which had their origins in the latter part of the Precambrian. Body styles that evolved in the early Cambrian largely served as blueprints for more modern forms. Only a few new radical body plans have appeared since. All known animal phyla that readily fossilized came into existence during the 60-million-year Cambrian period. Then in the late Cambrian, the number of new classes of animals suddenly declined.

The crustaceans appearing at the very base of the Cambrian soon became the dominant invertebrates. They occupied shallow waters near the shores of ancient seas, which flooded inland to provide extensive continental margins. The period is noted for a well-known group of invertebrates called trilobites (Fig. 6), which are the beloved of fossil hunters. They were primitive, oval-shaped crustaceans, ancestors of the horseshoe crab, their only remaining direct descendent. The giant trilobite *Paradoxides* was truly a paradox, for it reached nearly 2 feet in length while most other trilobites measured less than 4 inches long.

The trilobites diversified into about 10,000 species before declining and becoming extinct after a highly successful habitation spanning some 340 million years. The population of trilobites peaked around 520 million years ago, when they accounted for about two-thirds of all marine species. Since trilobites were so widespread and lived throughout the Paleozoic, their fossils are important markers for dating rocks of this era.

By about 475 million years ago, the number of trilobite species abruptly declined to one-third following the rise of mollusks, corals, and other stationary filter feeders. One dominant group of trilobites called the Ibex fauna vanished at the end-Ordovician mass extinction 440 million years ago possibly due to widespread glaciation. They were succeeded by the Whiterock fauna, which tripled their genera and sailed through the extinction practically unscathed. Eventually, the trilobites left the near shore for the off shore prob-

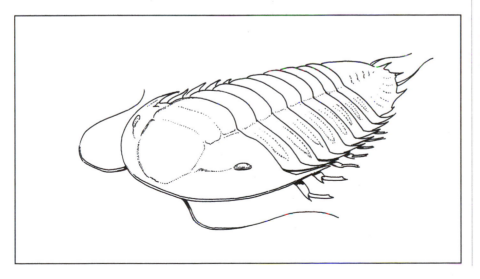

Figure 6 *Trilobites first appeared in abundance in the Cambrian and became extinct at the close of the Paleozoic.*

ably in response to significant environmental changes. The demise of the trilobites also appears to follow the arrival of the jawed fish.

THE AGE OF FISH

By the middle Paleozoic, the warm seas promoted the evolutionary development of marine species, including the first appearance of the ammonoids, which became one of the most highly successful creatures in the sea. The marine fauna included all forms that existed in the lower Paleozoic except the brachiopods, which decreased in number and types, and the trilobites, which rapidly declined and became extinct at the end of the era.

The first vertebrates appeared in the ocean about 530 million years ago and lacked jaws, fins, or true vertebrae. Among the oldest known vertebrates were primitive, jawless fish called agnathans (Fig. 7). They initially appeared about 470 million years ago and became most numerous between 420 and 390 million years ago. The presence of freshwater jawless fish suggests that unicellular plants, including red and green algae, which served as food, inhabited lakes and streams by this time.

The earliest fish were small mud grubbers and sea squirts that lacked jaws and teeth and were probably poor swimmers that occupied shallow waters. They had a flexible rod similar to cartilage running down their back instead of a bony spine typical of most vertebrates. They set the stage for the evolution of the backbone, one of the most important developments in the animal body plan.

The early jawless fish were generally small, about the size of a minnow, and heavily armored with bony plates protecting a rounded head. The rest of the body was covered with thin scales that ended near a narrow tail. Although

Figure 7 *The jawless fish agnathans were the progenitors of today's fish.*

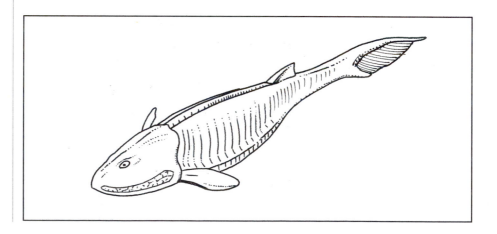

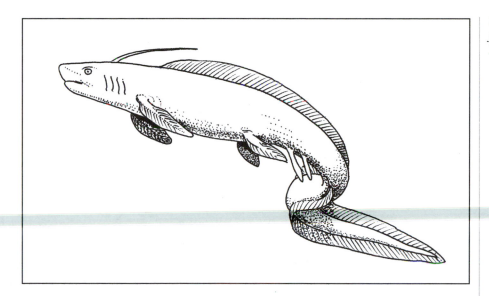

Figure 8 *An ancient freshwater shark called* Xenocanthus.

well protected from invertebrate predators, the added weight kept the fish mostly on the ocean bottom. There it sifted mud for food particles and expelled the waste products through slits on both sides of the throat that eventually evolved into gills.

Remarkably well-preserved remains of primitive fish were discovered in the mountains of Bolivia, which was inundated by the sea in the early to middle Paleozoic. Originally, fossil evidence was scant and fragmentary. Little was known of the fish's appearance or about its evolutionary history. Earlier descriptions dismissed the discovery as a headless, tailless mass of scales and plates. Sometimes, the head was confused with the tail, giving the remains the dubious title of "backward fossil."

The fossil record reveals so many and varied kinds of fish that paleontologists have difficulty classifying them all. Some earlier fish were surprisingly large, up to 18 inches long and 6 inches wide. Gradually, the protofish acquired jaws and teeth, the bony plates evolved into scales, lateral fins developed on both sides of the lower body for stability, and air bladders maintained buoyancy. Thus, for the first time, vertebrates skillfully propelled themselves through the sea. Fish soon became masters of the deep. They progressed from having rough scales, asymmetrical tails, and cartilage in their skeletons to having flexible scales, powerful advanced fins and tails, and all-bone skeletons, much like fish today.

The sharks were among the most successful fish. An ancient freshwater shark called Xenocanthus (Fig. 8) had a back fin that stretched from head to tail, allowing it to slither through the water like an aquatic snake. Closely

Figure 9 The rays flew through the sea on extended pectoral fins.

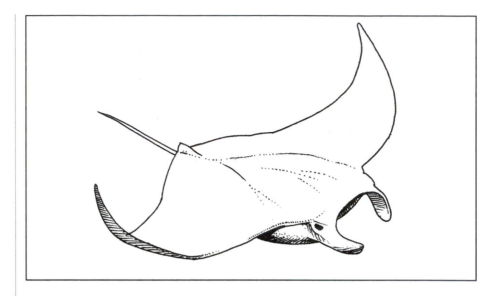

related to the sharks are the rays (Fig. 9), with flattened bodies, pectoral fins enlarged into wings up to 20 feet across, and a tail reduced to a thin, whiplike appendage. The rays literally fly through the sea, scooping plankton into their mouths.

The extinct placoderms (Fig. 10) were ferocious giants that reached 30 or more feet in length and preyed on smaller fish. They had well-developed articulated jaws and thick armor plating around the head that extended over and behind the jaws. Paired fins of an unusual construction were present on the lower body.

The coelacanth was thought to have gone extinct along with the dinosaurs 65 million years ago but was rediscovered in the deep, cold waters

Figure 10 The fierce placoderms were heavily armored and extended 30 feet in length.

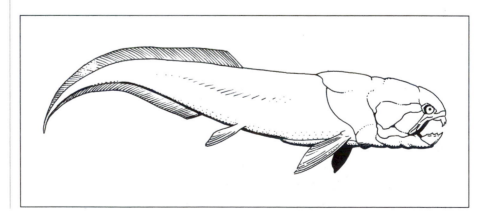

of the Indian Ocean in 1938. The coelacanth changed little from its primitive ancestors, which evolved in the Devonian seas some 370 million years ago, giving the fish the title of *living fossil*. It has a fleshy tail, a large set of forward fins behind the gills, powerful square jaws with inch-long, razor-sharp teeth, and heavily armored scales. It crawls along on the deep ocean floor on stout fins in a manner similar to its ancestors when they emerged from the sea to populate the land.

THE AGE OF AMPHIBIANS

The vertebrates finally set foot onto dry land about 360 million years ago, some 100 million years after terrestrial plants came ashore. By 400 million years ago, 2-inch-long plants with water-conducting vessels shared the land with a variety of arthropods. By the middle Paleozoic, stiff competition in the sea encouraged amphibious fish to make short forays onto the beach to prey on abundant crustaceans and insects. Freshwater invertebrates and fish inhabited the lakes and streams, while large cockroaches, giant dragonflies with 3-foot wingspans, sparrow-sized mayflies, and other monstrous insects occupied the forests.

The insects' exaggerated size supports the notion that oxygen comprised as much as 35 percent of the atmosphere. Species took advantage of the oxygen-rich air by growing large. The 100-million-year oxygen boost might also have given primitive amphibians an opportunity to develop their lungs and establish themselves full-time on land. However, by around 250 million years ago, the oxygen upheaval suddenly ended. Levels plummeted to about 15 percent. This was possibly due to massive volcanic eruptions, whose environmental impacts could have led to the mass extinctions at the end of the Paleozoic.

Insects are by far the largest living group of arthropods and can easily claim the title of the world's most prosperous creatures. Insects and plants have been at war with each other for more than 300 million years. The fiercest battles have been played out in the tropics, where hordes of hungry pests attack vegetation. Ever since animals left the sea and took up a lifestyle on dry land, insects and other arthropods have ruled the planet. The thorax, or midsection, supports three pairs of legs and typically two pairs of wings, which helped launch insects to their great success.

The insect body is usually covered with an exoskeleton made of chitin, similar to cellulose. In some groups, the exoskeleton was composed of calcite or calcium phosphate, which enhanced the insects' chances of fossilization. Generally, however, because of their delicate bodies, insects did not fossilize well unless entombed in amber, which is petrified tree sap that trapped unwary specimens and preserved them for the fossil record.

The crustaceans were probably the first invertebrates to crawl out of the sea and populate the land. These ancient arthropods emerged from the ocean soon after plants began to colonize the continents. The oldest known land-adapted animals were centipedes and tiny, spiderlike arachnids about the size of a flea. The arachnids are air-breathing crustaceans that include spiders, scorpions, daddy longlegs, ticks, and mites. The early terrestrial communities probably supported small, plant-eating arthropods that served as prey for these predatory animals.

The early crustaceans were segmented creatures, ancestors of today's millipedes, that walked on as many as 100 pairs of legs. They remained near the shore and eventually moved farther inland along with the mosses and lichens. With no competitors and abundant food supplies, some species evolved into the first terrestrial giants.

During this time, all major plant phyla were present. The higher land plants were firmly established on the previously barren continents. Not long after the plants crept ashore, the land was sprawling with lush forests. Due to fierce competition in the marine environment, creatures crawled out of the sea to dine on a plentiful food supply. Eventually, the vertebrates, the highest form of marine animal life, dominating all other species, left their homes in the sea to establish a permanent residence on land.

The ancestors of the amphibians are thought to be the crossopterygians and lungfish, from which all land vertebrates descended. By about 335 million years ago, the amphibious fish gave rise to the earliest amphibians. The amphibians began to dominate the land by the middle Devonian and were especially attracted to the great swamps. When the climate chilled and glaciers spread over the continents at the end of the period, the first reptiles emerged and began to displace the amphibians as the dominant land vertebrates.

THE AGE OF REPTILES

At the beginning of the Mesozoic, major groups of terrestrial vertebrates appeared on the scene, including the ancestors of reptiles, dinosaurs, and mammals. The direct ancestors of modern animals emerged, with the possible exception of the true birds. The first flight-worthy birds appeared in the Jurassic period and shared the skies with flying reptiles called pterosaurs (Fig. 11), which dominated the airways for 120 million years. A particular group of animals that excelled during the exceptionally mild climate was the reptiles. They occupied nearly every corner of the globe. Besides conquering the land, some

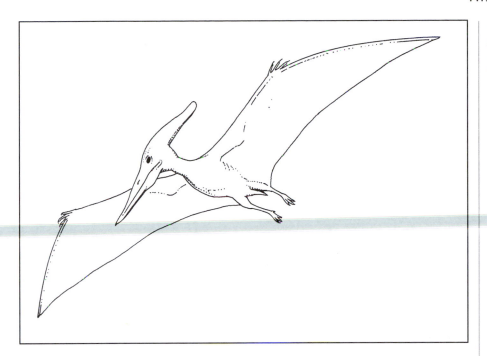

species returned to the sea to compete with the fish, while others took to the air to become the greatest aviators ever known.

When the Carboniferous swamps dried out and were largely replaced with deserts, the amphibians gave way to their reptilian cousins (Fig. 12), which were better suited to drier conditions. The once prominent amphibians declined significantly due to their preference for an aquatic life. The reptiles eventually became the dominant land-dwelling animals of the Mesozoic. The generally warm climate of the era was beneficial to the reptiles, enabling them to colonize the continents rapidly.

The age of reptiles spanned roughly 200 million years and gave rise to some 20 orders of reptilian families. At the end of the Triassic period about 210 million years ago, however, large families of terrestrial animals began dying off in record numbers. The extinction, spanning less than a million years, killed off nearly half the reptile families. The die out of species radically changed the character of life on Earth, paving the way for the dinosaurs.

THE AGE OF DINOSAURS

The dinosaurs arose to become the dominant terrestrial species for 170 million years. They suppressed the rise of other creatures. This includes the mam-

Figure 12a, b, c
Evolutionary stages in the development of the reptile. Amphibious fish (top), amphibian (middle), reptile (bottom).

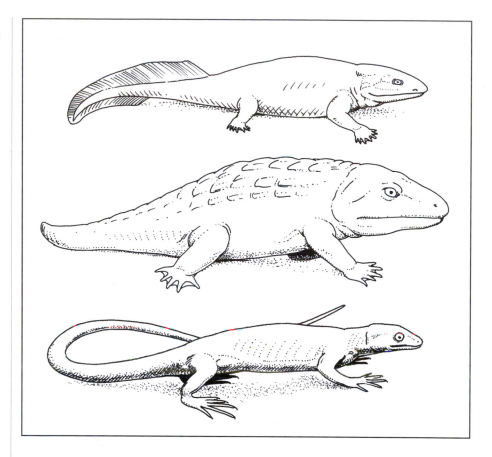

mals, which were small, rodentlike animals that sparsely populated the land and were hardly noticed since they did not directly compete with the dinosaurs or make much of a meal. The success of the dinosaurs is exemplified by their extensive range. They occupied a wide variety of habitats and dominated all other forms of land-dwelling animals. Indeed, had the dinosaurs not become extinct, mammals would never have achieved dominion over the Earth, and humans would not be here. The dinosaurs would have continued to stifle further advancement of the mammals, which would have remained small and inconspicuous.

The dinosaurs diversified into hundreds of species by about 200 million years ago during the Jurassic period. At that time, they reached their maximum size, becoming the largest terrestrial animals that ever lived. Widespread plant growth and coal formation suggest a warm, moist climate with excellent growing conditions that led to many giant dinosaur species. Perhaps the most celebrated of these was the huge bipedal brute *Tyrannosaurus rex,* which fully deserved the title "ruling reptile." The familiar triceratops with its three distinct horns rumbled through the countryside in huge herds, resembling modern galloping rhinoceroses. Nearby, chomping on lush vegetation were graceful apathosaurs (Fig. 13), some of the largest dinosaurs that ever lived.

Several families of large dinosaurs became extinct at the end of the Jurassic about 135 million years ago. Following the extinction, the population of small animals exploded as species occupied habitats vacated by the large dinosaurs. Most of the surviving species were small land dwellers and aquatic animals confined to freshwater lakes and marshes. Some species even had what appears to be feathers, suggesting that birds might have evolved from small, feathered dinosaurs. The feathers probably functioned as insulation or as courtship display similar to the activity of some modern birds.

At the end of the Cretaceous period 65 million years ago, another extinction struck down the dinosaurs, only this time none survived. Many small nondinosaur species endured the extinction possibly due to their large populations and abilities to find places to hide. This is especially true for the mammals, which radiated outward to fill habitats vacated by the dinosaurs. Because the dinosaurs represented the largest group of animals, their departure left the stage wide open for the mammalian invasion.

THE AGE OF MAMMALS

Mammals originated roughly the same time as the dinosaurs and coexisted with them for about 160 million years, as suggested by tiny, fossilized mam-

Figure 13 Dinosaurs such as apathosaur reached their maximum size during the Jurassic period.

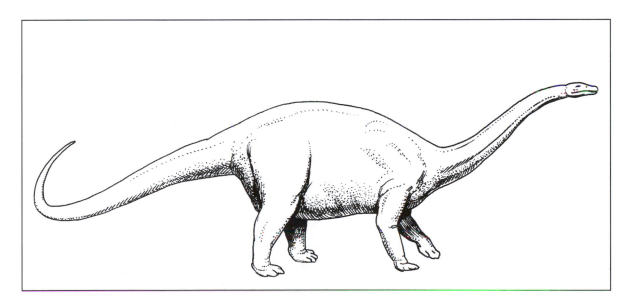

mal teeth. One of the oldest mammalian skeletons ever uncovered is that of a 140-million-year-old symmetrodont, which appears to be a missing link between egg-laying and therian (live-birth) mammals. A similar link was a bizarre, rat-sized animal dating back to at least 120 million years ago that walked on mammalian front legs and splayed reptilian hind legs. Around 100 million years ago, mammals began to branch into novel forms as new environments became available after the breakup of the supercontinent Pangaea.

Primitive mammals were relegated to the role of shrewlike nocturnal hunters of insects until the dinosaurs disappeared. With the dinosaurs no longer a threat, the mammals were set free to conquer the world, radiating into dazzling arrays of new species. The small mammals became recipients of daylight niches and eventually evolved into larger animals, many of which became evolutionary dead ends.

Of the 30 or so orders of mammals living in the early Cenozoic, only half lived in the proceeding Cretaceous and nearly two-thirds are still alive. In less than 10 million years following the extinction of the dinosaurs, all 18 modern mammal orders were established, including the primate order where humans belong. About 37 million years ago, a sharp extinction took many archaic mammal species. These were large, peculiar-looking animals that apparently could not adapt to changing climatic conditions. Henceforth, most truly modern mammals began to evolve. Mammalian evolution was not a gradual process, however, but progressed in steps. The early Tertiary was characterized by an evolutionary lag, as though the world had not yet awakened from the dinosaur extinction. However, by about 55 million years ago, mammals began to diversify rapidly.

Australia is home to many strange egg-laying mammals called monotremes, including the spiny anteater and platypus. These should rightfully be classified as surviving mammal-like reptiles (refer to chapter 3 on mammal-like reptiles). Monotremes arose more than 200 million years ago. The higher mammals including the placentals and marsupials diverged from a common ancestor in the early Cretaceous. The placental mammals, which produced fully developed infants, originated in Asia and migrated to North America. Marsupials are primitive mammals that suckle their tiny premature infants in belly pouches. They originated in North America about 100 million years ago, migrated to South America, crossed over to Antarctica, and finally landed in Australia (Fig. 14) before the continents drifted apart.

Presently, 13 of the world's 16 marsupial families reside in Australia alone. The Australian marsupials include kangaroos, wombats, and bandicoots, whereas opossums and related animals occupy other parts of the world. The largest marsupial fossil is that of diprotodon (Fig. 15), which was about the size of a rhinoceros. The giant kangaroos disappeared shortly after the arrival of early humans, possible ancestors of the Aborigines, that invaded the continent perhaps as early as 60,000 years ago.

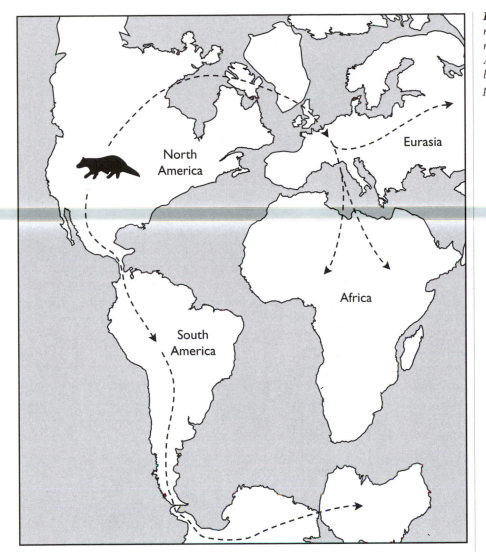

The first primates were about the size of a mouse and lived some 60 million years ago. The prosimians originally appeared about 50 million years ago. They gave rise to the anthropoids, the higher primates and ancestors of humans and apes. The primate family tree then split into two branches. Monkeys were on one limb. The great apes, including the hominoids—the predecessors of humans—were on the other.

About 30 million years ago, the precursors of apes lived in the dense tropical rain forests of Egypt, which has since become mostly desert. Between 25 million and 10 million years ago, these apelike ancestors migrated out of Africa to Europe and Asia (Fig. 16). From about 9 to 4 million years ago, the

Figure 15 *Diprotodon was the world's largest marsupial.*

fossil record jumps from a hominidlike but mainly ape form to the true hominids and the human line of ascension. During this time, much of Africa entered a period of cooler, drier climates and retreating forests, offering many evolutionary challenges to our human ancestors.

Figure 16 *The migration routes of African hominids.*

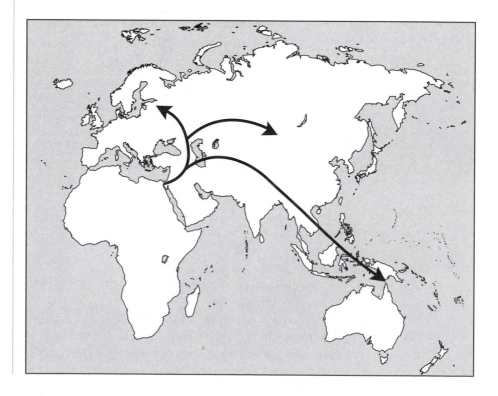

Figure 17 *Our human ancestry probably began with* Australopithecus.

(Photo courtesy National Museums of Canada)

An early hominid species called *Australopithecus* (Fig. 17) first appeared in Africa about 4 million years ago. Two or more lines of australopithecene existed simultaneously and survived practically unchanged for more than a million years. After a lengthy period of apparent stability, all but one line became extinct. This was possibly due to a changing climate or habitat. The gracile (thin-boned) australopithecenes apparently produced the *Homo* line of hominids.

A larger-brained hominid called *Homo habilis* evolved in Africa about 2 million years ago. It appears to have been a transition between primitive ape-like and humanlike hominids. It was much like *Australopithecus* but with a significantly larger brain. Around 1.8 million years ago, *Homo habilis* disappeared from Africa and was replaced by *Homo erectus*. It is widely accepted as human and appears to have evolved from *Homo habilis*. *Homo erectus* occupied southern and eastern Asia, where the species lived until about 200,000 years ago.

The Neandertals were a primitive species of *Homo sapiens* and first appeared about 300,000 years ago. Neandertals ranged over most of western Europe and central Asia, and they extended as far north as the Arctic Ocean. They thrived in these regions until about 30,000 years ago, apparently becoming extinct over a period of perhaps 5,000 years. The sudden disappearance of the Neandertals after more than 100,000 years of prosperity might have resulted from extinction by the hand of a more advanced human species known as Cro-Magnon, our direct human ancestors. For the past 25,000 years or so, *Homo sapiens* has had the Earth to itself, free of competition from other members of the hominid family.

After learning about the history of life, the next chapter will examine more closely the species of the sea from simplest protozoans to advanced marine tetrapods.

2

SEA LIFE
MARINE LIFE FORMS

On an uncharted seamount off the Bahamas, biologists of the Smithsonian Institution using a deep-sea submersible made a remarkable discovery that could open a new realm of oceanography. The scientists found a totally new and unexpected plant at a depth of about 900 feet, deeper than any previously known marine plant larger than a microbe. Previously, marine biologists thought the deepest that plants could survive was 600 feet because very little sunlight penetrates below that depth.

The deep-sea plant is a variety of purple algae. However, it has a unique structure consisting of heavily calcified lateral walls and very thin upper and lower walls that enable it to make the most of the feeble sunlight. The cells are stacked on top of one another like a stack of cans at a grocery store. The discovery will force scientists to rethink the role that such algae play in the productivity of the oceans, marine food chains, sedimentary processes, and reef building. This chapter explores the ocean depths in search of other remarkable life forms.

PROTOZOANS

Protozoans are the most primitive of animals and have survived throughout three-quarters of Earth history. They are often classified into the kingdom Protistae, which includes all single-celled plants and animals with a nucleus. In the obscured past, few distinctions existed between early plants and animals. They shared many similar characteristics. The protozoans are also classified into the animal kingdom. Indeed, the term literally means *beginning animals.*

The first group of organisms to evolve a nucleus was the protists. Some varieties formed large colonies, whereas most lived independently. The entire body is composed of a single cell containing living protoplasm enclosed within a membrane. Present-day single-celled organisms have not changed significantly from ancient fossils. However, most were soft bodied and did not fossilize well. They obtained energy by ingesting food particles or by photosynthesis. Reproduction is by using unicellular gametes that do not form embryos. The major groups of protists include algae, diatoms, dinoflagellates, foraminifers, fusulinids, and radiolarians (Fig. 18).

The foraminifers were microscopic protozoans. Their skeletons were composed of calcium carbonate, which preserved much of the record of the behavior of the ocean and climate. Most lived on the bottom of shallow seas, with a few floating forms existing. Their remains are found in both shallow- and deep-water deposits. The fusulinids were large, complex protozoans that resembled grains of wheat, ranging from microscopic size to giants up to 3 inches in length.

Among the earliest protists were microorganisms that built stromatolite structures. Ancestors of blue-green algae built these concentrically layered mounds, resembling cabbage heads, by cementing sediment grains together using a gluelike substance secreted from their bodies. As with modern stromatolites, the ancient stromatolite colonies grew in the intertidal zone. Their height, which was as much as 30 feet, indicated the height of the tides during their lifetime.

The organisms ranged from the Precambrian to the present, although many did not become well established until the Cambrian or later. Because of their lack of a hard shell, the amoebas and paramecia did not fossilize well. The radiolarians, which are well represented in the fossil record, usually have shells made of silica. These shells display remarkably intricate designs, including needlelike, rounded, or an open network structure of delicate beauty (Fig. 18).

The protists have a unique form of travel. Some unicellular animals move about with a thrashing tail, called a flagellum. It resembles a filamentous bacterium that joined with the single-celled animal in symbiosis for mutual benefit. Other cells have tiny hairlike appendages, called cilia, that help them travel by rhythmically beating the water. The amoeba has an unusual form of

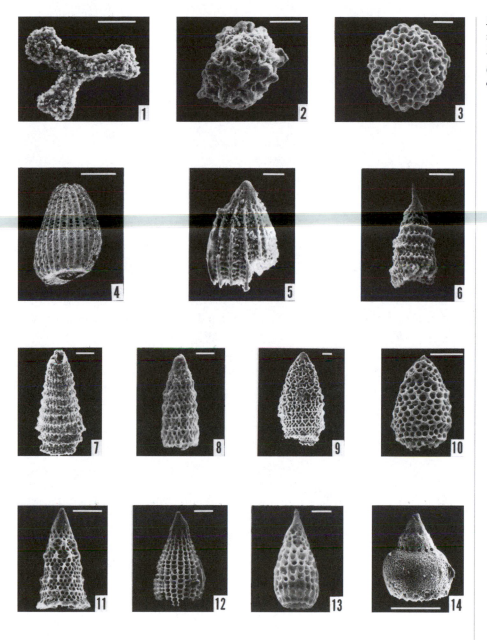

Figure 18 *Late Jurassic radiolarians, Chulitna District, Alaska.*

(Photo by D. L. Jones, courtesy USGS)

transportation: extending fingerlike protrusions outward from the main body and flowing into them.

Some varieties of protists performed a major function by building massive formations of limestone, which entombed other species as fossils. When the tiny organisms died, their shells fell to the ocean floor like a constant rain. The shifting of these sediments by storms and undersea currents buried dead marine organisms that were not eaten by scavengers. A calcite ooze was then formed, which eventually hardened into limestone, preserving trapped species for all time.

METAZOANS

In the latter part of the Proterozoic, around 600 million years ago, individual cells joined to form multicellular animals called metazoans. The metazoans gave rise to ever more complex organisms, which were ancestral to all marine life today. The first metazoans were a loose organization of cells united for a common purpose such as locomotion, feeding, and protection. If cells became separated from the main body, they could exist on their own until they regrouped or grew separately into mature species.

The first metazoans were probably composed of a large aggregate of cells, each with its own flagellum. The cells congregated into a small, hollow sphere. Their flagellum rhythmically beat the water to propel the tiny creature around. Some metazoans were literally turned inside out and attached themselves to the ocean floor. They had openings to the outside. The flagella, now on the inside, produced a flow of water that carried food particles in and wastes out. These were probably the forerunners of the sponges.

The sponges were the most primitive multicellular animals and thrived on the ocean floor just before the beginning of the Cambrian. They possessed various shapes and sizes and often grew in thickets on the ocean floor. Some early species became the first giants (Fig. 19), growing 10 feet or more across.

Figure 19 Early sponges were among the first giants on the ocean floor.

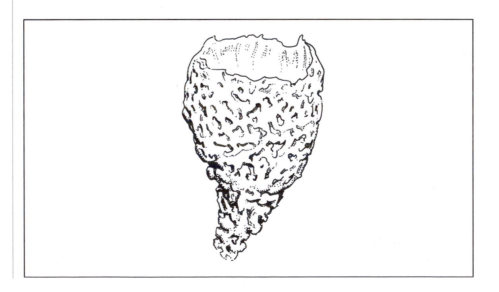

Figure 20 *A helmet jellyfish under the ice at McMurdo Sound.*

(Photo by W. R. Curtsinger, courtesy U.S. Navy)

The body consisted of three weak tissue layers whose cells could survive independently if separated from the main body and grow individually into mature sponges. Hence, sponges lack regulatory genes that tell developing embryos where to put new body parts as in complex animals.

Some groups have an internal skeleton of rigid, interlocking spicules composed of calcite or silica. One group had tiny, glassy spikes for spicules, which gave the exterior a rough texture unlike their smoother relatives used today in the bathtub. The so-called glass sponges consisted of glasslike fibers of silica intricately arranged to form a beautiful network. These hard skeletal structures are generally the only sponge parts preserved as fossils. Sponges ranged from the Precambrian to the present. However, microfossils of sponge spicules did not become abundant until the Cambrian. The great success of the sponges, which like other animals extract silica directly from seawater to construct their skeletons, explains why today's ocean is largely depleted of this mineral.

The jellyfish (Fig. 20) were an evolutionary step above the sponges. They have two layers of cells separated by a gelatinous substance, giving the saucerlike body a means of support. Unlike sponges, the cells of the jellyfish are incapable of independent survival if separated from the main body. A primitive nervous system links the cells, which contract in unison. These became the first simple muscles used for locomotion. Because they lacked hard body parts, jellyfish are rare as fossils and are usually preserved only as carbonized films or impressions.

The annelids are segmented worms with a body characterized by a repetition of similar parts in a long series. They include marine worms, earthworms, flatworms, and leeches. The primitive segmented worms developed muscles and other rudimentary organs, including sense organs and a central nervous system to process the information. The coelomic, or hollow-bodied, worms adapted to burrowing into the bottom sediments and might have led to more advanced animals.

The annelids ranged from the upper Precambrian to the present. Their fossils are rare and consist mostly of tubes, tiny teeth and jaws, and a preponderance of fossilized tracks, trails, and burrows. Marine worms burrowed into the bottom sediments or were attached to the seabed, living in tubes composed of calcite or aragonite. The tubes were almost straight or irregularly winding and attached to a solid object such as a rock, a shell, or coral. Early forms of marine flatworms grew very large, nearly 3 feet long.

Soft-bodied marine animals living in the Silurian period some 430 million years ago included a host of worms and bizarre arthropods. More enigmatic yet was a small, stubby-legged worm called a lobopod. A large variety of wormlike creatures lived in the Cambrian and apparently evolved into higher forms of animal life. The burrowing, shelled marine invertebrates, which evolved late in the Cambrian, often disrupted the corpses of soft-bodied animals buried in the sediment before fossilization could occur. This made the wormlike creatures extremely rare in the fossil record.

Graptolites (Fig. 21) were colonies of cupped organisms that resembled stems and leaves of plants but were actually animals. They clung to the seafloor

Figure 21 *Graptolites appeared to have gone extinct in the late Carboniferous.*

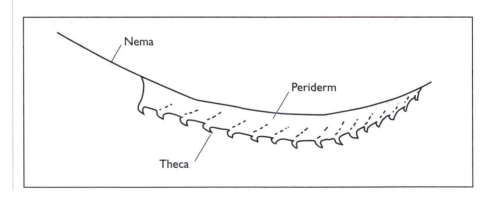

like small shrubs, floated freely near the surface by having tiny sawlike forms, or attached themselves to seaweed. Large numbers of graptolites buried in the bottom mud contributed to the formation of organic-rich black shales, making them important markers for correlating rock units of the lower Paleozoic. Graptolites appeared to have gone extinct in the late Carboniferous about 300 million years ago. However, the discovery of living pterbranchs, possible modern counterparts of graptolites, suggest these might be living fossils.

SHELL-LESS INVERTEBRATES

The coelenterates, from Greek meaning "gut," are among the lowest forms of animal life. The body is sac shaped with a mouth surrounded by tentacles. Most coelenterates are radially symmetrical, with body parts radiating outward from a central point. Primitive, radially symmetrical animals have just two types of cells, the ectoderm and endoderm. In contrast, the bilaterally symmetrical animals also have a mesoderm and a distinct gut. During early cell division, called cleavage of bilateral animals, the fertilized egg forms two, then four cells. Each then produces several small cells. Many types have two stages of development. One is a sessile benthic polyp attached to the seabed by its base with tentacles and mouth directed upward. The other is a pelagic, umbrella-shaped medusa or jellyfish-like phase with tentacles and mouth directed downward.

The coelenterates are well represented in the fossil record by the prolific corals. These are cupped animals attached to the ocean floor that exhibit a wide variety of skeletal forms (Fig. 22). Many corals declined in the late Paleozoic and were replaced by sponges and algae when the seas receded. Corals are soft-bodied animals that live in individual skeletal cups or tubes composed of calcium carbonate called thecae. The coral polyp is essentially a contractible sac crowned with tentacles (Fig. 23). The tentacles surround a mouthlike opening and are tipped with poisonous stingers to attack prey swimming nearby. The polyps extend their tentacles to feed at night and withdraw into their thecae during the day or at low tide to prevent drying in the sun.

Corals began constructing reefs in the early Paleozoic, forming barrier islands and island chains. Reef-building corals created the foundations for spectacular underwater edifices that cover about three-quarters of a million square miles and house about a quarter of all marine species. The corals diverged into two basic lineages prior to their ability to build calcified skeletons. This suggests they might have evolved a reef building mechanism twice in geologic history.

The corals live symbiotically with zooxanthellae algae within their tissues. The algae consume the coral's waste products and, in return, provide

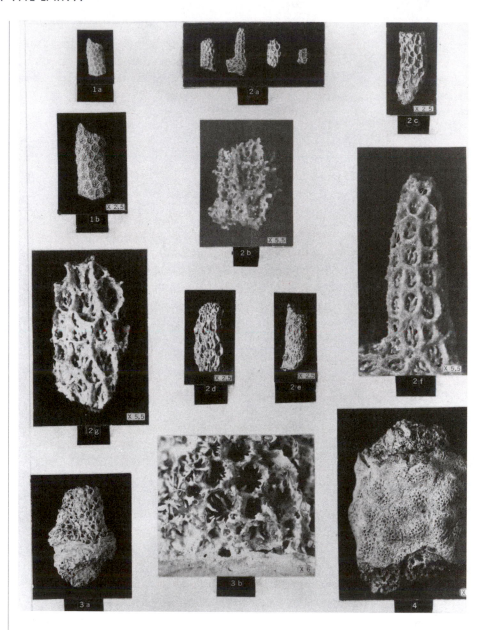

Figure 22 *Fossil corals from Bikini Atoll, Marshall Islands.*

(Photo by J. W. Wells, courtesy USGS)

organic materials that nourish the polyp. Some coral species receive 60 percent of their food from algae. Because the algae require sunlight for photosynthesis, corals are restricted to warm, shallow seas generally less than 100 feet deep and normally thrive at temperatures between 25 and 29 degrees Celsius.

Bryozoans (Fig. 24) resemble corals on a smaller scale but are more closely related to brachiopods. They consist of microscopic individuals living

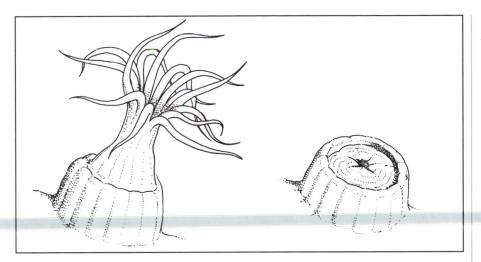

Figure 23 *The coral polyp is a contractible animal that lives in an individual skeletal cup.*

in small colonies up to several inches across, giving the ocean floor a mosslike appearance, which is why they are often called moss animals. Bryozoans are retractable creatures, encased in calcareous vaselike structures into which they retreat for safety when threatened. Living species occupy seas at various depths, with certain rare members adapted to life in freshwater.

A single free-moving larva bryozoan establishes a new colony by fixing onto a solid object. It then grows into many individuals by a process of budding, which is the production of outgrowths. The polyp has a circle of ciliated

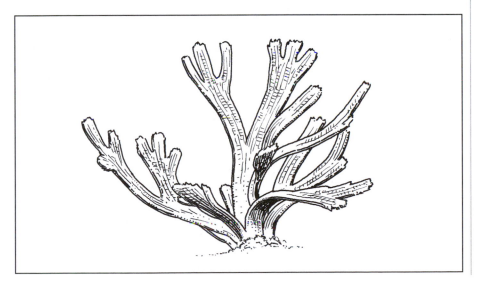

Figure 24 *The extinct bryozoans were major Paleozoic reef builders.*

tentacles, forming a sort of net around the mouth and used for filtering microscopic food floating by. The tentacles rhythmically beat back and forth, producing water currents that aid in capturing food. This is digested in a U-shaped gut. Wastes are expelled outside the tentacles just below the mouth.

Fossil bryozoans are abundant in Paleozoic formations, especially those of the American Midwest and Rocky Mountains. Bryozoan species are identified by the complex structure of their skeletons, which aids in delineating specific geologic periods. Bryozoans have been very abundant, ranging from the Ordovician to the present. Their fossils are highly useful for making rock correlations. Because of their small size, bryozoans make ideal microfossils for dating oil well cuttings.

Bryozoan fossils commonly occur in sedimentary rocks, particularly when covering the bedding surfaces of rocks. They resemble modern descendants. Some larger groups possibly contributed to Paleozoic reef building, producing extensive limestone formations. The fossils are most abundant in limestone and less plentiful in shales and sandstones. Often, a delicate outline of bryozoans can be seen encrusting fossil shells of aquatic animals, stones, and other hard bodies.

The echinoderms, from Greek meaning "spiny skin," are perhaps the strangest animals preserved in the fossil record of the early Paleozoic. They exhibit both bilateral symmetry and fivefold radial symmetry, with arms radiating outward from the center of the body. This unusual assembly makes echinoderms highly unique among the more complex animals. They are the only organisms possessing a water vascular system of internal canals for operating a series of tube feet called podia used for locomotion, feeding, and respiration.

Echinoderms were among the most prolific animals of the middle Paleozoic. The great success of the echinoderms is illustrated by the inclusion of more classes than any phylum both extinct and extant. The major classes of living echinoderms include crinoids, sea cucumbers, sea urchins, brittle stars, and starfish (Fig. 25). Sea urchins graze on coral reefs by scraping the rock surface as they feed, thereby contributing to the erosion of the reef structure.

The echinoderms were protected by exoskeletons consisting of numerous calcite plates. Starfish were common and left fossils in the Ordovician rocks of the central and eastern United States. Their skeletons were comprised of tiny silicate or calcite plates that were not rigidly joined and usually disintegrated on death, making whole starfish fossils rare. The sea cucumbers with large tube feet modified into tentacles had skeletons composed of isolated plates that are occasionally common as fossils.

The most successful echinoderms were the crinoids, commonly called "sea lilies" because they resembled flowers atop long stalks anchored to the ocean floor by rootlike appendages. The crinoid stalk consisted of perhaps 100 or more calcite disks, called columnals, and grew upward of 10 feet or more

Figure 25 *A starfish off Point Loma, San Diego, California.*

(Photo by R. Outwater, courtesy U.S. Navy)

in length. On weathered limestone outcrops, fossil crinoid stalks often bear resemblance to long strings of beads. A cup called a calyx perched on the stalk housed the digestive and reproductive organs. The animal strained food particles from passing water currents by using five feathery arms that extended from the calyx, giving the crinoid its flowerlike appearance. Some crinoids were also free-swimming types. The crinoids became the dominant echinoderms of the middle and upper Paleozoic and have many living relatives. The extinct Paleozoic crinoids and their blastoid cousins, whose calyxes resembled rose buds, made excellent fossils.

The echinoids are a class of echinoderms that include sea urchins, heart urchins, and sand dollars. Their skeletons are composed of limy plates and are characteristically spiny, spherical, or radially symmetrical. Some more advanced forms were elongated and bilaterally symmetrical. The sea urchins lived mostly among rocks encrusted with algae, upon which they fed. Unfortunately, due to its exposure, this environment was not very conducive to fossilization. Similarly, the familiar sand dollars that occasionally wash up on beaches rarely occur in the fossil record.

SHELLED INVERTEBRATES

The brachiopods (Fig. 26), also called lampshells because of their resemblance to ancient oil lamps, were once the most abundant and diverse organisms.

More than 30,000 species are cataloged in the fossil record. The appearance of large numbers of brachiopod fossils in rock formations indicates seas of moderate-to-shallow depth. Brachiopods had two saucerlike shells called valves fitted face to face that opened and closed by using simple muscles. More advanced species, including living brachiopods called articulates, had ribbed shells with interlocking teeth that opened and closed along a hinge line.

On the inside of the valves is a membrane called a mantle. The mantle encloses a large central cavity that holds the lophophore, which functions in food gathering. Projecting from a hole in the valve is a muscular stalk called a pedicel by which the animal is attached to the seabed. The structure of the valves aids in identifying the various brachiopod species. The shells come in a variety of forms, including ovoid, globular, hemispherical, flattened, convex-concave, or irregular. The surface is smooth or ornamented with ribs, grooves, or spines. Growth lines and other structures show changes in form and habit that offer clues to brachiopod history.

Brachiopods were fixed to the ocean floor by rootlike appendages and filtered food particles through opened shells that closed to protect the animals against predators. Brachiopod shells are often confused with those of clams, which are bivalves belonging to the more advanced mollusks. Clam shells are typically left- and right-handed in relation to the body and are mirror images of each other, with each valve being asymmetrical down the centerline.

The prolific brachiopods ranged from the Cambrian to the present. However, they were most abundant in the Paleozoic and to a lesser extent in the Mesozoic. Many brachiopods are excellent index fossils for correlating rock formations throughout the world. They are important as guide fossils and are used to date many Paleozoic rocks.

Mollusks are a highly diverse group that left the most impressive fossil record of all marine animals (Fig. 27). They make up the second largest of the 21 basic animal groups. The phylum is so diverse that finding common features among its members is often difficult. The mollusks were well represented in the middle Paleozoic. The first appearance of freshwater clams suggests that aquatic invertebrates had conquered the land by this time.

The mollusks became the most important shelled invertebrates of the Mesozoic seas, with some 70,000 distinct species living today. The warm climate of the Mesozoic influenced the growth of giant animals in the ocean as well as on land. Clams grew to 3 feet wide, squids were 65 feet long and weighed over a ton, and crinoids reached 60 feet in length.

The mollusk shell is an ever-growing, one-piece coiled structure for most species and a two-part shell for clams and oysters. Mollusks have a large

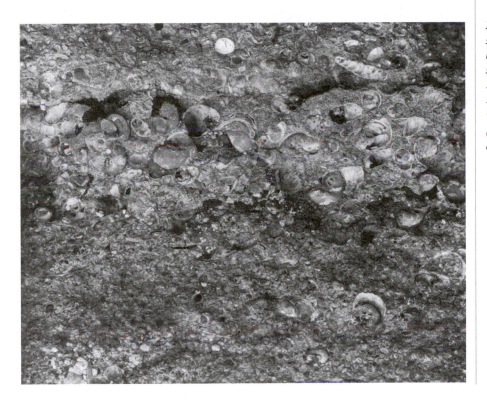

Figure 27 Molds and shells of mollusks on highly fossiliferous sandstone of the Glens Ferry Formation on Deadman Creek, Elmore County, Idaho.

(Photo by H. E. Malde, courtesy USGS)

muscular foot used for creeping and burrowing or modified into tentacles used for seizing prey. The clams are generally burrowers, although many species are attached to the ocean floor. The shell consists of two valves that hang down on either side of the body. Except for scallops and oysters, they mirror each other.

The three major groups of mollusks are snails, clams, and cephalopods. Snails and slugs make up the largest group and ranged throughout the Phanerozoic. The cephalopods include squid, cuttlefish, octopus, nautilus, and the extinct ammonite. They travel by jet propulsion, sucking water into a cylindrical cavity from openings on each side of the head and expelling it under pressure through a funnel-like appendage.

The nautiloids and ammonoids first appeared about 400 million years ago. For more than 300 million years, giant mollusks called ammonites were among the swiftest and most successful cephalopods, evolving into some 10,000 species. The two forms most well preserved in the fossil record were those with coiled shells (Fig. 28) up to 7 feet wide and more awkward types with straight shells growing to 12 feet in length.

The arthropods are the largest group of living organisms, with roughly a million species or about 80 percent of all known animals. The segmentation of the arthropod body suggests a relationship to the annelid worms. Paired, jointed limbs generally present on most segments were modified for sensing, feeding, walking, and reproduction. The giant three-foot-long arthropods found in the middle Cambrian Burgess Shale Formation of western Canada represent one of the largest of all invertebrates of that time.

The crustaceans are primarily aquatic arthropods and include shrimp, lobsters, crabs, and barnacles. The ostracods, or mussel shrimp, are small crus-

Figure 28 *The highly diverse ammonite shells helped date sedimentary strata.*

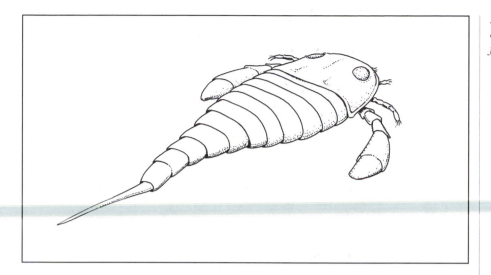

Figure 29 *The extinct
eurypterids grew to six
feet in length.*

taceans found in both marine and freshwater environments. Their fossils are useful for correlating rocks from the early Paleozoic onward, which make them particularly important to geologists. The extinct giant sea scorpion eurypterid (Fig. 29) lived from the early to the late Paleozoic and grew to 6 feet in length, terrorizing the ocean floor with immense pincers. Its descendants were among the first creatures to come ashore to prey on crustaceans.

Among the first and best known of the ancient arthropods were the trilobites (Fig. 30). These were aquatic crustaceans that lived mostly on the seafloor and were among the earliest known organisms to grow hard shells. The trilobite body was segmented into three lobes (hence its name). It consisted of a central axial lobe containing the essential organs and two side, or pleural, lobes. Trilobite fossils are often found with their bodies curled up, possibly as a form of protection. Trilobites shed their exoskeletons as they grew. Therefore, an individual could leave several fossils, which explains why whole specimens are rare. During molting, a suture opened across the head, and the trilobite simply fell out of its exoskeleton. Sometimes, though, a clean suture failed to open, and the animal had to wiggle its way out. Either way, it was still vulnerable to predators until its new skeleton hardened.

Many trilobite fossils show rounded bite scars predominantly on the right side of the body. Predators might have attacked from this direction possibly because when the trilobite curled up to protect itself, it exposed its right side. If the trilobite had a vital organ on its left side and was attacked there, it stood a good chance of being eaten, thereby leaving no fossil. However, if attacked on the right side, the trilobite had a better chance of entering the fossil record, albeit with a chunk bitten out of its body.

Figure 30 *Trilobites of the Carrara Formation in the southern Great Basin, California.*

(Photo by A. R. Palmer, courtesy USGS)

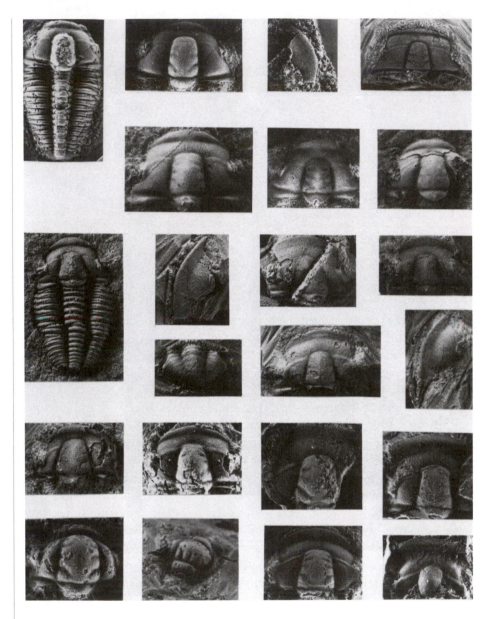

MARINE VERTEBRATES

Conodonts (Fig. 31) are tiny, jawlike appendages possibly belonging to the most primitive of vertebrates. They commonly occur in Paleozoic marine rocks from 520 million to 200 million years ago and are important markers for dating sediments of this era. They are among the most baffling of all fossils and have puzzled paleontologists since the 1800s. At that time, paleontologists began finding these isolated toothlike objects in rocks from the late Cambrian through the Triassic periods. Conodonts show their greatest diversity during the Devonian and are important for long-range rock correlations of that period.

The conodonts are thought to be bony appendages of an unusual, soft-bodied animal resembling a hagfish. However, the shape of the missing creature remained unclear until 1983, when paleontologists discovered toothlike pieces at the forward end of an eel-shaped fossil from Scotland. Furthermore, the presence of distinctive eye muscles not known in invertebrates pushed the vertebrate fossil record as far back as the Cambrian.

The most primitive of chordates, which include the vertebrates, was a small, fishlike oddity called amphioxus. Although the animal did not have a backbone, it nonetheless is placed in direct line to the vertebrates. The earliest vertebrates lacked jaws, paired fins on either side of the body, or true vertebrae and shared many characteristics of modern lampreys. The origin of the vertebrates led to the evolution of one of the most important novelties—namely the head. It was packed with paired sensory organs, a complex three-part brain, and many other features missing in invertebrates.

The vertebrates first appeared in the geologic record about 520 million years ago. They had internal skeletons made of bone or cartilage, one of life's most significant advancements. An internal skeleton enabled the wide dispersal of free-swimming species into a variety of environments. Also, the vertebrate skeleton was light, strong, and flexible, with efficient muscle attachments. The skeleton grew along with the animal. Invertebrates, supported by external skeletons, were at a distinct disadvantage in terms of mobility and growth.

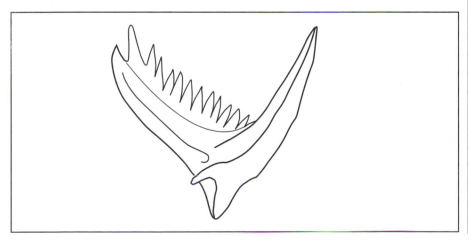

Figure 31 Conodonts became extinct in the Triassic.

Many animals such as crustaceans had to shed their shells as they grew, thereby making them vulnerable to predators.

The earliest vertebrates were probably wormlike creatures with a stiff, spinelike rod called a notochord running down the back to support organs and muscles, a system of nerves along the spine, and rows of muscles arranged in a banded pattern attached to the backbone. Rigid structures made of bone or cuticle acted as levers. Flexible joints efficiently translated muscle contractions into organized movements, such as rapid lateral flicks of the body to propel the animal through the water. A tail and fins eventually evolved to provide stabilization. The body became more streamlined and torpedo shaped for speed. With intense competition among the stationary and slow-moving invertebrates, any advancement in mobility was highly advantageous to the vertebrates. The widespread distribution of primitive fish fossils throughout the world suggests a long vertebrate record early in the Paleozoic.

Fish comprise over half the species of vertebrates, both living and extinct. The first protofish were jawless, generally small (about the size of a minnow), and heavily armored with bony plates. Although well protected from their invertebrate enemies, the added weight kept these fish mostly on the bottom, where they sifted mud for food particles. Gradually, the protofish acquired jaws with teeth, the bony plates gave way to scales, lateral fins developed on both sides of the lower body for stability, and air bladders were used for buoyancy.

A revolution in predation occurred with the development of jaws about 460 million years ago. Giant jawed vertebrates, some of which were monsters in their day, climbed to the very top of the food chain. One of these groups gave rise to land animals, emphasizing the great importance jaws played in vertebrate evolution. The development of jaws also improved fish respiration by supporting the gills. After a fish draws water into its mouth, it squeezes the gill arches to force the water over the gills at the back of the mouth. Blood vessels in the gills exchange oxygen and carbon dioxide as the water flows out the gill slits. The jaws had the advantage of clamping down on significantly large prey, allowing fish to become fierce predators. Primitive jawed fish might have even caused the demise of the trilobites, once spectacularly successful in the Cambrian seas.

The rise of fish in the Devonian contributed to the decline of less mobile invertebrate competitors. This culminated in an extinction that eliminated many tropical marine groups at the end of the period. When a mass extinction occurs, organisms that have evolved into a better adaptive form prior to the event are selected for survival, which is why certain species survive one major extinction after another. This is particularly true for marine species such as sharks. They originated in the Devonian around 400 million years ago and have withstood every mass extinction since.

Sharks breathe by drawing water in through the mouth, passing it over the gills, and expelling it through distinctive slits behind the head. Because the shark's body is heavier than water, it must continually swim or sink to the ocean bottom. Instead of skeletons made of bone as with most fish, those of sharks are composed of cartilage, a much more elastic and lighter material. However, cartilage does not fossilize well. Essentially, the only common shark remains are teeth found in marine sediments from the Devonian to the present.

Links between fish and terrestrial vertebrates were the lungfish, another living fossil still surviving today in Africa, Australia, and South America, all of which were continents of Gondwana. The crossopterygians were lobe finned, with the bones in their fins attached to the skeleton and arranged into primitive elements of a walking limb. They breathed by drawing air into primitive nostrils and lungs as well as by using gills. This placed them into the direct line of evolution from fish to land-living vertebrates.

The lobe-finned fish, which first appeared about 410 million years ago, had thick, rounded fins whose bones were crude forerunners of those in tetrapod (four-legged) limbs. Lobe-finned fish used gills for respiration but could also breathe with primitive lungs in oxygen-poor swamps or when stranded on dry land. Their descendants became the first advanced animals to populate the continents.

The likely ancestors of the amphibians were the crossopterygians, the predecessors of modern lungfish and the stem group from which all tetrapods descended. They grew up to 10 feet long and were lobe-finned, air-breathing fish with heavy, enamel-like scales and large teeth. An abundance of food swept onto the beaches during high tide might have enticed these fish to come ashore briefly. By digging in the sand for food and shelter, the crossopterygians strengthened their fins, which gradually evolved into walking legs.

The coelacanth, originating from the same evolutionary branch in direct line to land-dwelling vertebrates, used its stout fins to crawl along on the deep ocean floor. The fins were coordinated in a manner common in tetrapods, moving similarly to the legs of a lizard. The forward appendage on each side advanced in concert with the rear appendage on the opposite side. Such an adaptation would have eased the transition from sea to land.

MARINE TETRAPODS

Marine mammals known as cetaceans, including whales, porpoises, and dolphins, are among the most adaptable animals. The dolphins had reached the level of intelligence comparable to living species by 20 million years ago probably due to the stability of their ocean environment. Sea otters, seals, walruses, and manatees (Fig. 32) are not fully adapted to a continuous life at sea and have retained many of their terrestrial characteristics. The manatees, which have inhabited Florida waters for 45 million years, are rapidly becoming an endangered species.

The smallest cetaceans would appear to be in danger of hypothermia in the cold northern waters. With a normal mammalian metabolic heat supply, a small porpoise could not possibly keep warm. Yet despite their size, cetaceans maintain a normal internal temperature of about 98.6 degrees Fahrenheit, the same as for humans. They accomplish this feat with a basal metabolism rate about three times higher than that of other mammals of the same weight. As a result, they must consume three times more food and oxygen than their terrestrial counterparts.

The pinnipeds, meaning *fin footed,* are a group of marine mammals with four flippers. The three surviving forms include seals, sea lions, and walruses. The true seals (Fig. 33) without ears are thought to have evolved from weasel-like or otterlike forms. Sea lions and walruses, in contrast, are believed to have developed from bearlike forms. This type of dual development, called diphyletic evolution, makes originally dissimilar pinnipeds resemble each other simply because they have adapted to a life in water. The similarity in their flippers, however, suggests that all pinnipeds evolved from a single land-based mammal that entered the sea millions of years ago.

For well over a century, zoologists have pondered how evolution managed to craft such unique creatures as whales, which presented a mystery as vast as the animals themselves. The legless leviathans had evolved from mam-

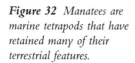

Figure 32 Manatees are marine tetrapods that have retained many of their terrestrial features.

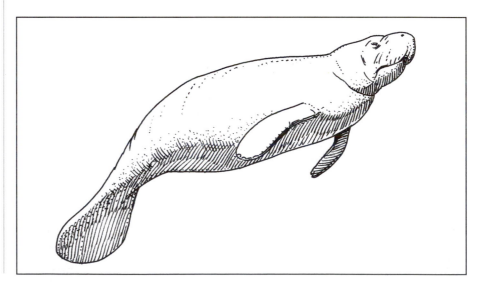

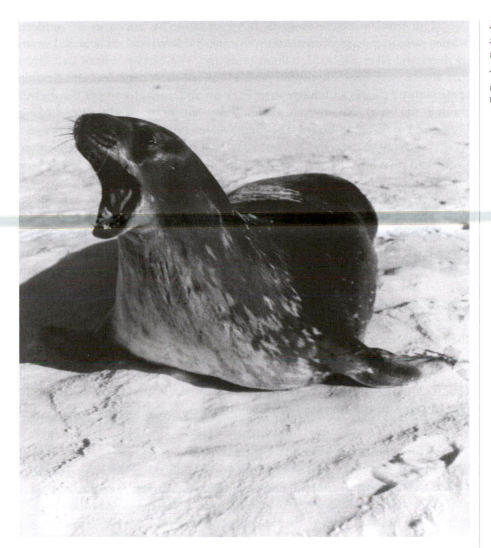

Figure 33 *A Weddell seal sunning on a pressure ridge near Scott Base, Antarctica.*

(Photo by Luethje, courtesy U.S. Navy)

mals known as ungulates, whose best-known characteristic is a set of hoofed feet. Whales adapted to a swimming, diving, and feeding mode of life that matches or surpasses fish and sharks. They might have gone through a seallike amphibious stage early in their evolution. Today, their closest relatives are the artiodactyls, or hoofed mammals with an even number of toes, such as cows, pigs, deer, camels, giraffes, and hippos. However, exactly where whales fit into the tree of hoofed mammals remains controversial. Genetic evidence suggests that whales and hippopotamuses are closely related. Both groups share particular aquatic adaptations, such as the ability to nurse their young and communicate underwater. The ancestor of both whales and hippos might have ventured into the sea as early as 55 million years ago.

rapidly. The early seaweeds were soft and nonresistant. Therefore, they did not fossilize well. A variety of fossil spores used for reproduction have been found in late Precambrian and Cambrian sediments, which suggests that complex sea plants were in existence. However, no other significant remains have been discovered. Even as late as the Ordovician, plant fossils appeared to be composed almost entirely of algae, which probably formed algal mats similar to those on seashores today. Once life crept ashore, however, plants quickly covered the Earth's surface with lush forests.

The first terrestrial plants were comprised of algae and early seaweeds residing just below the surface in the shallow waters of the intertidal zones. Primitive forms of lichen and moss lived on exposed surfaces. They were followed by tiny fernlike plants called psilophytes, or whisk ferns, the predecessors of trees. These simple plants lived semisubmerged in the intertidal zones, lacked root systems and leaves, and reproduced by casting spores into the sea for dispersal. The most complex land plants grew less than an inch tall and resembled an outdoor carpet covering the landscape.

By the middle Paleozoic, the terrestrial flora was plentiful and varied. The most significant evolutionary step was the development of a vascular stem to conduct water to a plant's extremities. The early club mosses, ferns, and horsetails were the first plants to utilize this water vascular system. The early complex land plants diverged into two major groups. One gave rise to the lycopods, including club mosses and scale trees (Fig. 34). The other spawned the gymnosperms, which were ancestral to many modern land plants. Eventually, as evolution steadily progressed, great coal forests of lycopods and gymnosperms spread across the continents. The woodlands included thick stands of seed ferns and true trees. These were gymnosperms with seeds and woody trunks. Forests transformed the planet with their green, lush vegetation some 370 million years ago.

Plants displayed increased diversity and complexity, including root systems, leaves, and reproductive organs employing seeds instead of spores. When the true leaves evolved, plants developed a variety of branching patterns to optimize leaf exposure to sunlight for maximum photosynthesis. As plants grew larger, they diverged from random branching to tiers of branches to achieve greater efficiency with minimum self-shading, similar to present-day pines.

The lycopods ruled the ancient swamps, towering as high as 130 feet. They were the first trees to develop true roots and leaves, which were generally small. Branches wound around in a spiral. Spores clung to modified leaves that evolved into primitive cones. The scale trees, so-named because the scarring on their trunks resembled large fish scales, grew to 100 feet or more high and were among the dominant trees of the late Paleozoic forests.

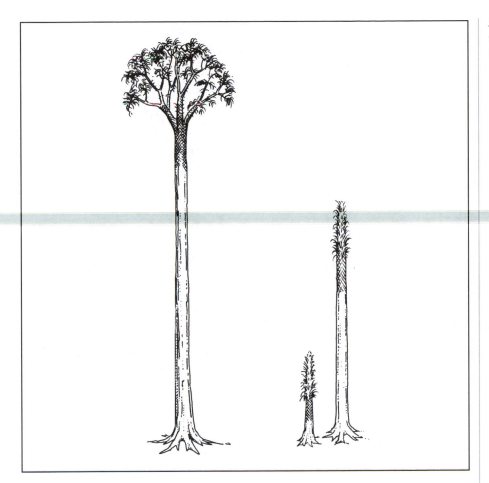

Figure 34 *The scale tree was one of many early trees.*

As the tropics grew more arid and the Carboniferous swamplands dried up, the climate change initiated a wave of extinctions that wiped out practically all lycopods at the beginning of the Permian 280 million years ago. Today, they exist only as small, grasslike plants in the tropics. As the climate grew wetter and the swamps reemerged, weedy plants called tree ferns dominated the Paleozoic wetlands (Fig. 35).

The mountainous landscape in the northern latitudes of Pangaea was covered by thick forests of primitive conifers, horsetails, and club mosses that grew to 30 feet tall. Much of the continental interior probably resembled a grassless rendition of the contemporary steppes of central Asia, with temperatures varying from very hot in summer to extremely cold in winter. Since grasses would not appear for another 100 million years or so, the scrubby landscape was dotted with bamboolike horsetails and bushy clumps of extinct seed ferns resembling present-day tree ferns.

Figure 35 Fossil leaves of the tree fern neuropteris, Fayette County, Pennsylvania.

(Photo by E. B. Harden, courtesy USGS)

The true ferns were the second most diverse group of living plants. They were particularly widespread during the Mesozoic and prospered well in the mild climates even in the higher latitudes. In contrast, today they survive only in the tropics. Some ancient ferns attained the heights of today's trees. The Permian seed fern glossopteris was particularly significant on the southern continent Gondwana. Its absence on the northern continent Laurasia signified the two large landmasses were separated by a vast gulf, thus preventing migration.

The gymnosperms, including cycads, ginkgos, and conifers, originated in the Permian and bore seeds lacking fruit coverings. The cycads resembled palm trees and were highly successful, ranging across all major continents. The ginkgo, of which the maidenhair tree in eastern China is the only living relative, might have been the oldest genus of seed plants. Also dominating the landscape were conifers up to 5 feet across and 100 feet long. Petrified trunks of conifers are well exposed in multiple tiers at Yellowstone National Park (Fig. 36).

The Mesozoic was a transition period for plants. They showed little resemblance at the beginning of the era to those at the end, when they more closely resembled present-day vegetation. A dramatic change in plant life

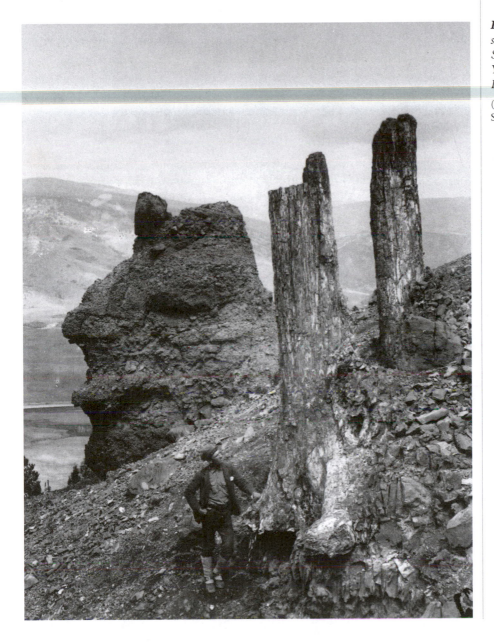

Figure 36 *Petrified tree stumps on the North Scarp of Specimen Ridge, Yellowstone National Park, Wyoming.*

(Courtesy National Park Service)

occurred about 100 million years ago with the introduction of the angiosperms. These are flowering plants that evolved alongside pollinating insects. Their sudden appearance and eventual domination over other plant life has remained a mystery. The earliest angiosperms were apparently large plants as tall as magnolia trees. However, fossils discovered in Australia suggest that the first angiosperms there and perhaps elsewhere were merely small, herblike plants.

Within a few million years after they burst onto the evolutionary scene, the efficient flowering plants crowded out the once dominant gymnosperms and ferns. They possessed water-conducting cells called vessel elements that enabled the advanced plants to cope with drought conditions. Before such vessels appeared, plants were restricted to moist areas such as the wet under-growth of rain forests. The angiosperms were widely distributed by the end of the Cretaceous. Today, they include about a quarter-million species of trees, shrubs, grasses, and herbs.

All major groups of modern plants were represented in the early Tertiary (Fig. 37). The angiosperms dominated the plant world. All modern families had evolved by about 25 million years ago. Grasses were the most significant angiosperms, providing food for hoofed mammals called ungulates. Their grazing habits evolved in response to the widespread availability of grasslands, which sparked the evolution of large herbivorous mammals and ferocious car-nivores to prey on them.

THE AMPHIBIAN INVASION

The plants had been greening the Earth for as long as 100 million years before the vertebrates finally began to set foot onto dry land some 360 million years ago. Prior to the amphibian invasion, freshwater invertebrates and fish had been inhabiting lakes and streams. The vertebrates had spent more than 160 million years underwater, with only a few short forays onto the land. Relatives of the lungfish lived in freshwater pools that dried out during seasonal droughts, requiring the fish to breathe with primitive lungs as they crawled to the safety of the nearest water hole.

The lobe-finned fish, which first appeared about 410 million years ago, had thick, rounded fins whose bones were crude forerunners of those in tetra-pod (four-legged) limbs. Lobe-finned fish breathed with gills. Unlike other fish, they also breathed with lungs in oxygen-poor swamps or when stranded on dry land. Their descendants became the first advanced animals to populate the continents.

The amphibious fish probably spent little time on shore and frequently returned to sea. Their primitive legs could support their body weight only for

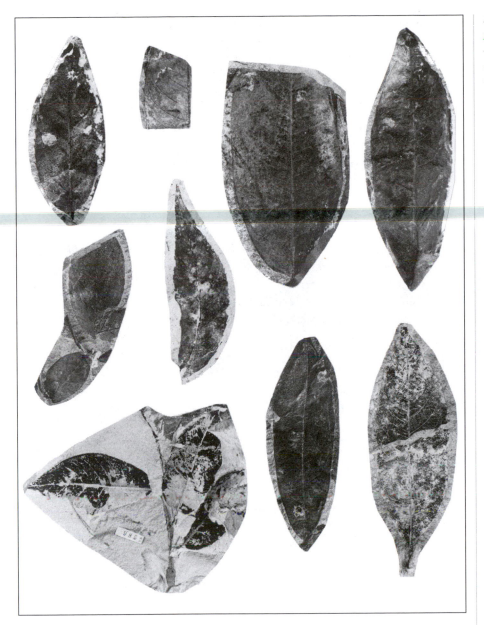

Figure 37 *Tertiary plants, Cook Inlet region, Alaska.*

(Photo by J. A. Wolfe, courtesy USGS)

short periods. However, as their limbs strengthened, the amphibious fish wandered farther inland, within easy access of nearby water sources such as swamps or streams where crustaceans and insects were abundant. The amphibious fish eventually evolved into the earliest amphibians, whose legacy is well documented in the fossil record. At no other time in geologic history had so many varied and unusual creatures inhabited the surface of the Earth.

Acanthostega, meaning *spine plate,* was the earliest known tetrapod. It had a salamanderlike body with large eyes on top of a flat head for spotting prey swimming above as it sat buried in the bottom mud. It sported eight toes on each front leg, perhaps the most primitive of walking limbs. The digits were sophisticated and multijointed. However, because they were attached to an insubstantial wrist, the legs were virtually useless for walking on the ground. The rest of the skeletal anatomy also suggests acanthostega could not have easily walked on land. Instead, it probably crawled around on the bottom of lagoons and used gills for respiration.

One of the earliest known amphibians was an ancient land vertebrate called ichthyostega (Fig. 38), meaning *fish plate.* It lived half the time in and half the time out of water. It was dog sized with a broad, flat, fishlike head and a small fin atop the tail, apparently used for swimming. It developed a sturdy rib cage to hold up its internal organs while on land and crawled around on primitive legs with seven toes on the hind limbs. Amphibians also possessed six and eight digits on their feet, indicating the evolution of early land vertebrates followed a flexible pattern of development. However, no terrestrial vertebrates evolved a foot with more than five true digits. Neither acanthostega nor ichthyostega could do much more than waddle around on land. Their upper arm bones had a broad, blobby shape ill suited for walking. Their hind limbs splayed out to the side and could not have easily held up the body.

The tetrapods branched into two groups around 330 million years ago. One line led to amphibians and the other to reptiles, dinosaurs, birds, and mammals. Some amphibians had strong, toothy jaws and resembled giant salamanders up to 5 feet in length. A 2-foot-long amphibian with armadillolike plates rooted in the soil for worms and snails.

Although the legs of the amphibians were well developed for walking on dry land, the animals apparently spent most of their time in rivers and

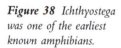

Figure 38 Ichthyostega was one of the earliest known amphibians.

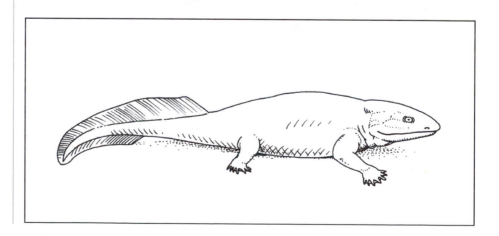

swamps. They depended on accessible sources of water to moisten their skins as well as for respiration and reproduction. They reproduced like fish, laying small, shell-less eggs. After hatching, the juveniles maintained an aquatic, fish-like existence, using gills for respiration. As they matured, the young amphibians metamorphosed into air-breathing, four-footed adults.

The weak legs of the early amphibians could hardly keep their squat bodies off the ground, making them slow and ungainly. The amphibian tracks are generally broad with a short stride. The animal walked with a clumsy gait. Running to attack prey or escape predators was simply not possible. Therefore, to succeed as hunters without requiring speed or agility, the amphibians developed a unique whiplike tongue that lashed out at insects and flicked them into the mouth. This successful adaptation enabled the amphibians to populate the land rapidly.

Their semiaquatic lifestyle, however, led to the eventual decline of the amphibians when the great swamps began to dry out toward the end of the Paleozoic. Populations of amphibians continued to fall during the Mesozoic, with all large, flat-headed species going extinct. The group thereafter was represented by the more familiar salamanders, toads, and frogs. Although the amphibians did not achieve complete dominion over the land, their cousins the reptiles became the undisputed rulers of the world.

COLD-BLOODED REPTILES

The first reptiles emerged some 300 million years ago. Within a short time span, they became the leading form of animal life, occupying land, sea, and air. The reptilian foot was a vast improvement over the amphibian mode of transportation, indicating reptiles were better suited for continuously living on dry land. Their advanced foot design included changes in the form of the digits, the development of a thumblike fifth digit, and the appearance of claws.

The tracks of some reptiles narrowed, and the stride lengthened. Many reptiles maintained a four-footed walking gait and reared up on their hind legs when running down prey. The body pivoted at the hips, and a long tail counterbalanced the nearly erect trunk. This stance freed the forelimbs for attacking small animals and for other useful tasks.

The reptilian body is covered with scales that retain bodily fluids, allowing the animal to live entirely on dry land. Another major advancement was the reptilian mode of reproduction by laying eggs with hard, watertight shells. Reptiles belong to a group known as amniotes. These are animals with complex eggs that evolved from the amphibians and also include birds and mammals. As with fish and amphibians, reptiles are cold-blooded and must draw heat from their environment. A high body temperature is as important to reptiles as it is to mammals to achieve maximum metabolic efficiency. The unusually warm climate of the Mesozoic probably contributed substantially to the success of the reptiles.

Reptiles require only a fraction of the food needed by mammals to survive, however, because mammals metabolize most of their calories to maintain a high body temperature. The total energy consumption of mammals is 10 to 30 times and the oxygen intake is about 20 times that of reptiles of the same weight. Consequently, reptiles can live successfully in deserts and other desolate places. They can flourish on small amounts of food that would quickly starve a mammal of equal size. Small reptiles probably ate insects as modern lizards do.

All early reptiles were not small creatures. Moschops (Fig. 39) were up to 16 feet long and had thick skulls adapted for butting during mating season, much like the behavior of modern herd animals. They might have been preyed upon by packs of lycaenops, which were reptiles with doglike bodies and long canine teeth projecting from their mouths.

Phytosaurs (Fig. 40) were large, heavily armored predatory reptiles sporting sharp teeth. They resembled crocodiles, with short legs, long tails, and elongated snouts, but were not closely related to them. They evolved from the thecodonts, the same group that gave rise to crocodiles and dinosaurs. The phytosaurs thrived in the late Triassic, evolving quite rapidly. Apparently they did not survive beyond the end of the period.

Pterosaurs were perhaps the most spectacular reptiles that ever lived. They had wingspans of up to 40 feet, about the size of a small aircraft, and dominated the skies for 120 million years. Their wings were similar in con-

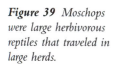

Figure 39 Moschops were large herbivorous reptiles that traveled in large herds.

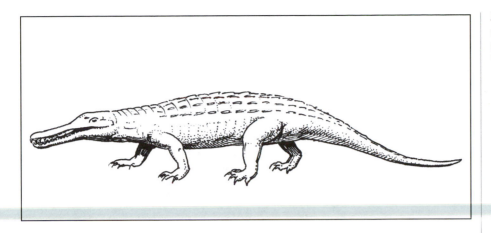

struction to a bat's wings and probably originated as a cooling mechanism that later evolved into wings when the advantages of flight became apparent. Pterosaurs probably spent much time aloft riding updrafts similar to modern birds of prey.

When the reptile was the leading form of animal life, a remarkable reptilian order (the Crocodylia) appeared in the fossil record. Members of this group adapted to life on dry land, a semiaquatic life, or a marine life with a sharklike tail, a streamlined head, and legs remolded into swimming paddles. Over the past 200 million years, the crocodilians diversified considerably and dispersed to all parts of the world, adapting to a wide variety of habitats. Around 100 million years ago, as with many land animals, crocodiles grew to giants as indicated by a 40-foot-long fossil found in southern Brazil. A fossil of a gaviallike monster from the lower Cretaceous in Niger, West Africa measured about 35 feet long. A monstrous 3-ton, 30-foot crocodile terrorized the Cretaceous swamps and might have preyed upon medium-sized dinosaurs.

The presence of crocodile fossils in the high latitudes of North America suggests a warm climate in this region during the Mesozoic. Along with dinosaurs and pterosaurs, the crocodilians belong to the subclass Archosauria, which literally means *ruling reptiles.* However, unlike the dinosaurs, they are the only surviving members to escape the great dying at the end of the Mesozoic.

MAMMAL-LIKE REPTILES

Mammal-like reptiles were animals in transition from reptiles to mammals. Fossilized bones of a mammal-like reptile with large down-pointing tusks called lystrosaurus (Fig. 41) were discovered in the Transantarctic Range of Antarctica. Its presence provides strong evidence that the continent had once

Figure 41 *Lystrosaurus was a mammal-like reptile that lived in Gondwana.*

Figure 41 *Lystrosaurus was a mammal-like reptile that lived in Gondwana.*

been joined with southern Africa and India. As ancestor to mammals, lystrosaurus was the most common vertebrate on land and was found throughout Gondwana. Mammal–like reptiles called dicynodonts also had two caninelike tusks and fed on small animals along riverbanks.

The first animals to depart from the basic reptilian stock were the pelycosaurs, which evolved about 300 million years ago. The distinguishing characteristics that set them apart from other reptiles were their larger body size and varied diet. The earliest pelycosaur predators could kill sizable prey, including large reptiles.

A pelycosaur called dimetrodon (Fig. 42) lived some 280 million years ago and grew to about 11 feet in length. Along its back, a tall dorsal sail composed of webs of membrane that were well supplied with blood stretched across thin, bony, protruding spines. The appendage probably provided a means of temperature control by absorbing sunlight during cold weather and radiating excess body heat when the weather was hot. This structure might have been a crude forerunner to the thermoregulatory system in mammals. The pelycosaurs eventually lost their sails as the climate turned warm and perhaps gained some degree of internal thermal control. They thrived for about 50 million years and were replaced by their descendants, the mammal–like reptiles called therapsids (Fig. 43).

The earlier therapsids retained many characteristics of the pelycosaurs, including legs well adapted for fast running. They ranged in size from as small as a mouse to as large as a hippopotamus. The early therapsids invaded the southern continents near the end of the Paleozoic when those lands were still recovering from glaciation. This suggests the animals were warm-blooded so they could withstand the cold. They therefore seemed to be well adapted for feeding and traveling through the snows of the cold winters. However, the

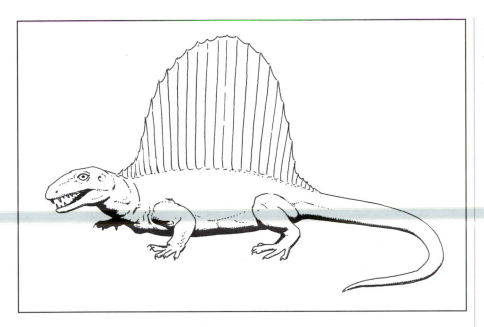

Figure 42 *A 300-million-year-old pelycosaur called dimetrodon had a huge sail to regulate its body temperature.*

therapsids were apparently too large to hibernate in winter, as evidenced by the absence of growth rings in their bones, similar to how tree rings indicate different growth rates during the seasons. The development of fur appeared in the more advanced therapsids as they migrated into colder climates. The therapsids might also have operated at lower body temperatures than most living mammals to conserve energy.

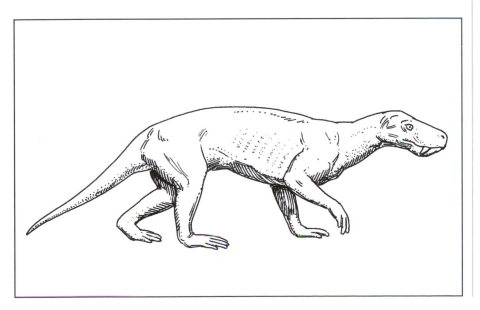

Figure 43 *Many therapsids were large, predatory, mammal-like reptiles.*

The therapsids supposedly reproduced like reptiles by laying eggs. They might have protected and incubated the eggs and fed their young. This behavior possibly led to longer egg retention and live births. The therapsids dominated animal life for more than 40 million years until the middle Triassic. Then for unknown reasons, they lost ground to the dinosaurs.

THE DINOSAUR DYNASTY

When the mammal-like reptiles dominated the land, dinosaurs represented only a small fraction of all animal life. Many reptile species living at the time of the early dinosaurs far outweighed them. However, within a few million years, dinosaurs rapidly became the dominant species, evolving from moderate-sized animals less than 20 feet long to the giants for which they are famous.

The oldest dinosaurs originated on the southern continent Gondwana when the region was recovering from late Paleozoic glaciation. They descended from the thecodonts (Fig. 44), the apparent common ancestors of crocodiles and birds. The earliest thecodonts were small-sized to medium-sized predators that lived during the Paleozoic-Mesozoic transition, which was an evolutionary heyday for land animals.

Figure 44 *Dinosaurs descended from the thecodonts.*

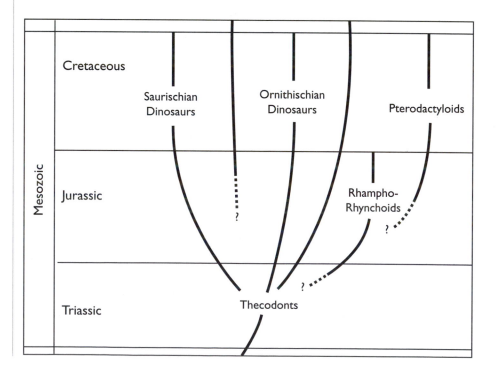

60

Figure 45 *The small plant eater camptosaur.*

(Courtesy National Park Service)

One group of thecodonts returned to the sea and became large, fish-eating, aquatic species. They included the phytosaurs, which died out in the Triassic, and the crocodilians, which remain successful today. Pterosaurs also descended from the thecodonts. The appearance of featherlike scales ostensibly used for insulation suggests that thecodonts were also the ancestors of birds. The protofeathers helped trap body heat or served as a colorful display for attracting mates as with modern birds. By about 200 million years ago, the thecodonts were replaced by the dinosaurs as the dominant terrestrial vertebrates.

Dinosaurs are classified as either sauropods or carnosaurs. Sauropods were long-necked herbivores. Carnosaurs were bipedaled carnivores that possibly hunted sauropods in packs. Camptosaurs (Fig. 45), ancestor to many later dinosaurs species, were herbivores up to 25 feet long. Protoceritops and heavily armed ankylosaurs (Fig. 46) were very common dinosaurs and ranged over wide areas of the world. Many large, bipedaled dinosaurs assumed similar appearances simply because of the need to balance themselves on two feet, which required a huge tail and small forearms. A two-legged sprint would also have been the fastest way to travel to run down prey.

Not all dinosaurs were giants, however. Many were no larger than most modern mammals. The smallest known dinosaur footprints are only about the size of a penny. Many early, small dinosaurs were the first animals to establish

Figure 46 *Ankylosaurs were common herbivorous dinosaurs.*

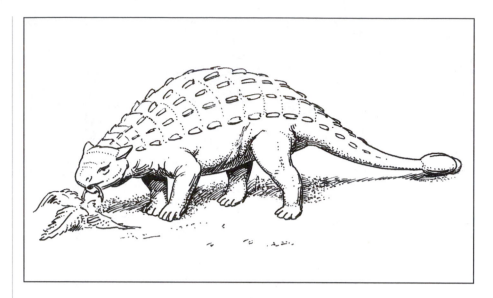

Figure 46 *Ankylosaurs were common herbivorous dinosaurs.*

a successful, permanent two-legged stance. Bipedalism freed the forelimbs for foraging and other tasks. The back legs and hips supported the entire weight of the animal, while a large tail counterbalanced the upper portions of the body. This made the dinosaur walk similarly to birds. Dinosaurs are therefore classified as ornithischians with a birdlike pelvis or as saurischians with a lizardlike pelvis. The ornithischians apparently arose from the same group of thecodont reptiles that were ancestral to crocodiles and birds.

Some large, bipedal dinosaurs later reverted to a four-footed stance as their weight increased. They eventually evolved into gigantic long-tailed, long-necked sauropods such as the apathosaurs, which are in the same family of dinosaurs as the brontosaurs. Others, such as *Tyrannosaurus rex* (Fig. 47), meaning *terrible lizard,* were perhaps the fiercest land carnivores of all time. They maintained a permanent two-legged stance with powerful hind legs, a muscular tail for counterbalance, and arms shortened to almost useless appendages. About 90 million years ago, when the continents became isolated and evolved separate fauna, *T. rex* began to roam the American West.

The carnivorous dinosaurs were cunning and aggressive, attacking prey with unusual voracity. The cranial capacity of some carnivores suggests they had relatively large brains and were fairly intelligent, able to react to a variety of environmental pressures. The velociraptors, meaning "speedy hunters" with their sharp claws and powerful jaws were vicious killing machines whose voracious appetites suggest they were warm-blooded.

During the Jurassic, dinosaurs attained their largest sizes and longest life spans. The biggest dinosaurs occupied Gondwana, which included all the

Large reptiles possess the power of almost unlimited growth and continue to grow throughout their lives. A large body helps cold-blooded animals maintain their body temperature for long periods. This makes the animal less susceptible to short-term temperature variations in the environment. Only the force of gravity kept the dinosaurs from growing larger than they did. When an animal doubles its size, the weight on its bones quadruples. The obvious exceptions were the dinosaurs that lived permanently in the sea. As with present-day whales, some of which are larger than the largest dinosaur that ever lived, the buoyancy of seawater kept the weight off their bones.

Among the largest dinosaur species were the apathosaurs and brachiosaurs, which lived around 100 million years ago. These gargantuan creatures fully deserve the title *thunder lizards.* They were sauropods with long, slender necks and tails, and the front legs were generally longer than the hind legs. Perhaps the tallest and heaviest dinosaur thus far discovered was the 80-ton ultrasaurus, which could tower above a five-story building. Seismosaurus, meaning "earth shaker," was the longest known dinosaur. It reached a length of more than 140 feet from its head, which was supported by a long, slender neck, to the tip of its even longer whiplike tail. Giganotosaurus even challenges *Tyrannosaurus rex* as the most ferocious terrestrial carnivore ever to live.

THE EARLY BIRDS

Birds first appeared in the fossil record about 150 million years ago, although some accounts push back their origin as much as 75 million years earlier. By about 135 million years ago, early birds began to diversify. They diverged into two lineages, one leading to archaic birds and the other leading to modern birds. The birds descended from the thecodonts, the common ancestors of dinosaurs and crocodiles. Consequently, they are often referred to as "glorified reptiles." Alternatively, they could have descended from bipedaled, meat-eating dinosaurs known as theropods, from Greek meaning "fierce foot." Indeed, birds are the only living relatives of dinosaurs. The skeletons of many small dinosaur species (Fig. 48) closely resembled those of birds, suggesting a direct descendence from dinosaurs.

Birds are warm-blooded to obtain the maximum metabolic efficiency needed for sustained flight but retain the reptilian mode of reproduction by laying eggs. The bones of some Cretaceous birds show growth rings, a feature common among cold-blooded reptiles. This suggests that early birds might not yet have developed fully warm-blooded bodies. The birds' ability to maintain high body temperatures has sparked a controversy over whether some dinosaur species with similar skeletons were warm-blooded as well.

southern continents. The generally warm climate and excellent growing conditions encouraged the growth of lush vegetation, including ferns and cycads to satisfy the insatiable appetites of the plant-eating dinosaurs. The huge size of the herbivores spurred the evolution of giant carnivorous dinosaurs to prey on them.

Many dinosaur species grew into giants for the same reasons elephants and other megaherbivores are so large. The majority of the large dinosaurs were herbivores, or plant eaters. Therefore, they had to consume huge quantities of coarse cellulose, which took a long time to digest. This required an oversized stomach for the fermentation process and, consequently, a large body to carry it around. Some species swallowed gizzard stones (like modern birds do) to grind the fibrous fronds into pulp. The rounded, polished stones were left in a heap where the dinosaur died. Deposits of these stones can be found on exposed Mesozoic sediments, especially in the American West.

Figure 4? Tyrannos *the greate* *that ever l*

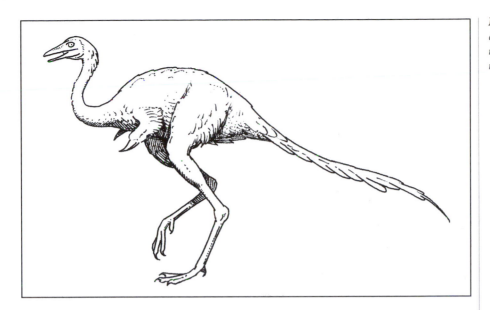

Figure 48 Many small dinosaurs, such as monoykus, were built much like birds.

Archaeopteryx (Fig. 49), from Greek meaning "ancient wing," is the earliest known fossil bird. It was about the size of a modern pigeon and appeared to be a species in transition between reptiles and true birds. However, unlike modern birds, it lacked a keeled sternum for the attachment of flight muscles. It was first thought to be a small dinosaur until fossils clearly showing impressions of feathers were discovered in 1863 from a unique limestone formation in Bavaria, Germany.

Although *Archaeopteryx* appeared to have all the necessary appendages for flight, it likely was a poor flyer, taking off only for short distances like today's domesticated birds. It probably became airborne either by running along the ground with its wings outstretched and then gliding for a brief moment or by leaping upward while flapping its wings to catch insects flying by. Their forebearers might have flapped their wings to increase running speed while escaping predators, thereby obtaining flight by pure accident. *Archaeopteryx* had teeth, claws, a long bony tail, and many skeletal features of small dinosaurs but lacked hollow bones to conserve weight. Its feathers were outgrowths of scales and probably originally functioned as insulation.

Some bird species retained their teeth until the late Cretaceous, roughly 70 million years ago. Claws on the forward edges of the wings might have helped the birds climb trees, from which they could launch themselves into the air. As birds mastered the skill of flight, they quickly radiated into a variety of environments. Their superior adaptability enabled them to compete successfully with the pterosaurs, possibly leading to the flying reptile's eventual decline.

Figure 49
Archaeopteryx *appears to be a link between reptiles and birds.*

Giant flightless land birds appeared early in the avian fossil record. Their wide distribution is further evidence for the existence of the supercontinent Pangaea since these birds had to travel on foot to get from one part of the world to another. After having been driven into the air by the dinosaurs, birds found life a lot easier on the ground once the dinosaurs disappeared because they had to expend a great deal of energy to stay in the air. Some birds also successfully adapted to a life in the ocean. Penguins (Fig. 50) are flightless birds that have taken to the sea and are well adapted for survival on the frozen wasteland of Antarctica. Certain diving ducks are specially equipped for "flying" underwater as well as in the air.

WARM-BLOODED MAMMALS

The first mammals were tiny, shrewlike creatures that appeared about 220 million years ago. They evolved from bulky, cold-blooded creatures that were themselves descendants of the reptiles. Mammals descended from the mammal-like reptiles, which were later driven into extinction by the dinosaurs about 160 million years ago. The mammals then began to branch into new forms following the breakup of Pangaea. The multituberculates were a distinctive mammalian species and probably the most interesting group of mammals that ever lived. They evolved about the same time as the dinosaurs and became extinct a little more than 30 million years ago, long after the dinosaurs disappeared.

The early mammals (Fig. 51) evolved over a period of more than 100 million years into the first therian (live-birth) mammals, the ancestors of all living marsupials and placentals. During this time, mammals progressed toward functioning better in a terrestrial environment. Teeth evolved from simple cones that were repeatedly replaced during an animal's lifetime into more complex forms replaced only once. However, the mammalian jaw and other parts of the skull still shared many similarities with reptiles. One mysterious group known as the triconodonts, ranging from 150 to 80 million years ago, were primitive protomammals. They were possible ancestors of the monotremes, which are represented today by the platypus and echidna. These lay eggs and walk with a reptilian, sprawling gait.

Australia is home to strange, egg-laying monotremes, which should rightfully be classified as surviving, mammal-like reptiles. The platypus has a ducklike bill, webbed feet, and a broad, flattened tail. The marsupials have belly pouches for incubating their tiny young after birth. Their ancestors originated in North America about 100 million years ago and migrated southward to Australia, using Antarctica as a land bridge.

The ancient mammals were forced into nocturnal lifestyles. This required the evolution of highly acute senses along with an enlarged brain to process the information. The archaic mammals developed a fourfold increase in relative brain size compared with the reptiles. Thereafter, they achieved no

Figure 50 A pair of Adélie penguins in Antarctica.

(Photo by M. Mullen, courtesy U.S. Navy)

Figure 51 *The extinct eurotamandua was an early anteater.*

substantial increase in brain size for at least 100 million years. This indicated an adaptation to a lengthy, stable ecological niche during the Mesozoic.

Following the extinction of the dinosaurs 65 million years ago, mammals became the recipients of daytime niches along with a preponderance of new sensory signals for the brain to organize as they competed with each other in a challenging environment. In less than 10 million years, all 18 modern orders of mammals were established. In addition, all orders of hoofed mammals emerged in full bloom after the dinosaur extinction.

Another fourfold increase in relative brain size occurred about 50 million years ago in response to adaptive radiation of mammals into new environments. During this time, rodents, the largest group of mammals, appeared in the fossil record. During the rest of the Cenozoic, mammalian brains gradually grew larger in proportion to their bodies. Intelligent activity is generally the key to mammalian success, implying a certain degree of freedom of action. With their superior brains, mammals could successfully compete with much stronger animals.

The number of mammalian genera rose to a high of 130 about 55 million years ago. Thereafter, the number of genera waxed and waned, dropping to as low as 60 and rising up to 120 presumably in response to climate change and migration. These fluctuations lasted millions of years, yet diversity always converged on an equilibrium of about 90 genera. Greater speciation resulting in stiffer competition caused higher extinction rates, which maintained mammalian genera at a constant number.

Mammals are warm-blooded, which gives them a tremendous advantage. A stable body temperature finely tuned to operate within a narrow thermal range provides a high rate of metabolism independent of the outside temperature. Therefore, the work output of the heart, lungs, and leg muscles increases enormously, allowing mammals to outperform and outendure reptiles. Mammals also have a coat of insulation, comprising an outer layer of fat and fur to prevent the escape of body heat during cold weather.

Other distinguishable features of mammals include four-chambered hearts, a single bone in the lower jaw, highly differentiated teeth, and three small ear bones that migrated from the jawbone backward as the brain grew larger to improve hearing greatly. Mammals have live births and possess mammary glands that provide a rich milk to suckle their young, which are generally born helpless. Mammals have the largest brains, capable of storing and retaining impressions. Therefore, they live by their wits, which explains their great success. They conquered land, sea, and air and are established, if only seasonally, in all parts of the world.

After examining life forms in all their seemingly limitless varieties, the next chapter will concentrate on mass extinctions in Earth history from the earliest Precambrian die-off of species to the loss of life following the Ice Age.

4

MASS EXTINCTIONS
THE DISAPPEARANCE OF SPECIES

This chapter chronicles the mass disappearance of species at various times in Earth history. Throughout the history of life, species have come and gone on geologic timescales so that those living today represent only a tiny fraction of the total. Of the estimated 4 billion species of plants and animals that have inhabited this planet in the geologic past, more than 99 percent have gone extinct. The vast majority of the Earth's fauna and flora lived during the Phanerozoic, the last 600 million years. This was a period of phenomenal growth as well as tragic episodes of mass extinction, each involving the loss of more than half the species living at the time.

Because of their significant impacts on life, major extinction events mark the boundaries between geologic periods. Each period is characterized by somewhat fewer profound changes in organisms as compared with the eras, which delineate boundaries of mass extinctions, proliferations, or rapid transformations of species. Beginning with the Cambrian explosion, an interval of astonishing diversity, five distinct mass extinctions marked the end of the Ordovician, Devonian, Permian, Triassic, and Cretaceous periods. The close of the Permian signaled the greatest extinction in the geologic record when more

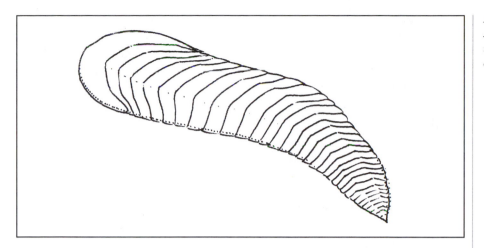

than 95 percent of the marine species and about 80 percent of the terrestrial species disappeared. In addition, five or more minor die outs during the Phanerozoic resulted in the disappearance of large numbers of other species.

PRECAMBRIAN EXTINCTIONS

The Earth experienced its first major glaciation about 2.2 billion years ago when ice sheets covered almost the entire planet, reaching as far as the tropics. The ice age caused a mass disappearance of primitive organisms attempting to evolve at this time. In the late Precambrian, around 800 million years ago, stromatolites, comprising concentric layers of sediment laid down by lime-secreting algae, suffered a marked decline in diversity. This coincided with the appearance of algae-eating animals including the earliest worms.

Late in the Precambrian, at least four major ice ages took hold of the planet between 850 and 600 million years ago. Around 670 million years ago, thick ice sheets spread over much of the landmass during perhaps the greatest period of glaciation the Earth has ever known. This is called the Varanger ice age, when massive ice sheets overran nearly half the continents for millions of years. At this time, all continents were assembled into a supercontinent named Rodinia, Russian for "motherland." Rodinia might have wandered over one of the poles and collected a thick sheet of ice. The ice age dealt a deathly blow to life in the ocean. Many simple organisms vanished during the world's first mass extinction, when animal life was still scarce. The late Precambrian extinction decimated the ocean's population of acritarchs, a community of planktonic algae that were among the first organisms to develop elaborate cells with nuclei.

Eventually, volcanic carbon dioxide spewing out from the Earth's interior over millions of years created a greenhouse effect powerful enough to break the ice's grip. Then the world began to thaw. When glaciation ended, life diversified in all directions. This resulted in an explosion of species that represented nearly every major group of marine organisms and set the stage for the evolutionary development of more modern life forms. The great diversity of animal life spurred the evolution of entirely new species, forever altering the composition of the Earth's biology. Many unique and bizarre creatures inhabited the Earth, whose fossil impressions left their mark in the rocks of the Ediacara Formation in South Australia (Fig. 52).

The Earth underwent many profound physical changes, prompting a rapid radiation of Ediacaran fauna. The supercontinent Rodinia was located on the equator. Intense volcanic and hydrothermal (hot-water) activity resulted in fundamental environmental changes. The seas overran the continents, producing wide continental margins. The increased marine habitat area prompted an explosion of new species, including large populations of widespread and diverse organisms.

Many of the unusual creatures that evolved during the late Precambrian adapted to highly unstable living conditions, including an increasing oxygen content. The abundant oxygen in the ocean permitted the evolution of large animals with vascular circulatory systems to supply cells with blood. The rising oxygen level led to the evolution of multicellular organisms, resulting in an exceptional diversity of animal life.

The first multicellular animals of sufficient size appeared in the fossil record roughly 600 million years ago. Prior to reaching a certain oxygen threshold, species remained small and simple because the low oxygen levels could not support the metabolic machinery of multicellular life. As the oxygen content continued to increase, the stage was set for intense evolutionary creativity. This led to the birth of nearly all modern groups of animals.

Overspecialization to a narrow range of environmental conditions, however, culminated in a major extinction of species at the end of the Precambrian due to the evolution of predators, establishing new predator-prey relationships. Species that survived the extinction differed markedly from their Ediacaran ancestors and participated in the greatest diversity of new organisms the world has ever known.

LOWER PALEOZOIC EXTINCTIONS

The early Paleozoic witnessed an explosion of new species, the likes of which have never been seen before. Then just 10 million years into the Cambrian period, a wave of extinctions decimated a large variety of newly evolved

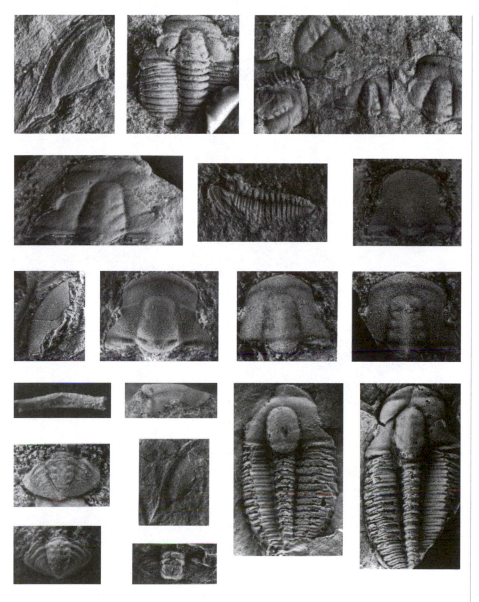

Figure 53 *Trilobite fossils of the Carrara Formation in the Southern Great Basin, California-Nevada.*

(Photo by A. R. Palmer, courtesy USGS)

organisms, eliminating more than 80 percent of all marine animal genera. The extinction is numbered among the worst in Earth history and coincided with a drop in sea levels. The die-offs led to the ascendancy of the trilobites (Fig. 53), which passed through the extinction relatively unscathed. They were among the first animals well protected with hard shells and became the dominant species for the next 100 million years.

The flora of the lower Paleozoic included cyanobacteria, red and green algae, and acritarchs, which supported early food chains. The vase-shaped

Figure 54
Archaeocyathans built the earliest limestone reefs.

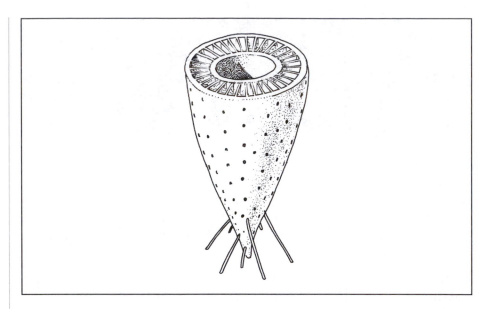

archaeocyathans (Fig. 54), from Greek meaning "ancient cup," built the earliest limestone reefs. They grew to about 2 inches tall and bore a resemblance to both sponges and corals. The archaeocyathans fell to extinction in the Cambrian period during a rapid rise of new competitors and a loss of habitat resulting from a drop in sea level. Since archaeocyathans were not closely related to any living group, they are classified into their own unique phylum. Many corals declined and were replaced by sponges and algae in the late Paleozoic due to the recession of the shallow seas where they had once thrived.

During the great Cambrian explosion, confined to a 5- to 10-million-year interval beginning about 530 million years ago, species diversity was at an all-time high. Many experimental organisms came into existence at this time, none of which have any modern counterparts. The soft-bodied animals of the Burgess Shale fauna, whose remains were discovered early in the 20th century in British Columbia, Canada, appeared soon after the emergence of complex life forms. The Burgess Shale invertebrates were surprisingly intricate. Some species might have been surviving Ediacaran fauna, most of which became extinct near the end of the Precambrian.

Most Burgess Shale faunas abruptly went extinct at the end of the Cambrian. Only a few archaic forms survived to the middle Devonian. Although many mass extinctions of marine organisms have occurred since the late Cambrian (Table 3), no fundamentally new body styles have appeared over the past 500 million years.

The movement of continents into the polar regions was thought to be responsible for a period of glaciation during the late Ordovician around 440

TABLE 3   RADIATION AND EXTINCTION OF SPECIES

Organism	Radiation	Extinction
Mammals	Paleocene	Pleistocene
Reptiles	Permian	Upper Cretaceous
Amphibians	Pennsylvanian	Permian-Triassic
Insects	Upper Paleozoic	Permian
Land plants	Devonian	Permian
Fish	Devonian	Pennsylvanian
Crinoids	Ordovician	Upper Permian
Trilobites	Cambrian	Carboniferous & Permian
Ammonoids	Devonian	Upper Cretaceous
Nautiloids	Ordovician	Mississippian
Brachiopods	Ordovician	Devonian & Carboniferous
Graptolites	Ordovician	Silurian & Devonian
Foraminifers	Silurian	Permian & Triassic
Marine invertebrates	Lower Paleozoic	Permian

million years ago. The study of magnetic orientations in rocks from many parts of the world indicates the positions of the continents relative to the magnetic poles at various times in geologic history. The paleomagnetic studies in Africa were very curious, however. The northern part of the continent was placed directly over the South Pole during the Ordovician, which led to widespread glaciation.

The glaciation struck during the invasion of land plants, which might have removed carbon dioxide, an important greenhouse gas, from the atmosphere. The frigid conditions resulted in a mass extinction that eliminated some 100 families of marine animals. Most victims were tropical faunas sensitive to extreme fluctuations in the environment. Among the losers were a large number of trilobite species. Prior to the extinction, trilobites accounted for about two-thirds of all animal species but only a third thereafter. Despite the excellent climate and the extreme success of the trilobites in the lower Paleozoic, they began to decline rapidly in the Silurian, with numerous species succumbing to extinction at the end of the period.

The tabulate corals, comprising closely packed polygonal or rounded calcite cups or thecae, were another group that became extinct at the end of the Paleozoic. The rugose, or horn, corals, so-named because of their typical

Figure 55 *The extinct
tetracorals were major reef-
building corals.*

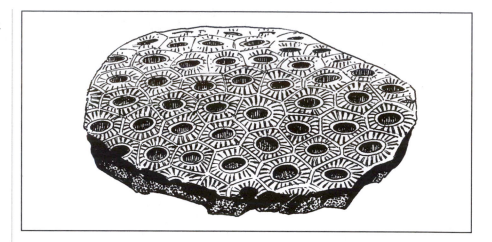

hornlike shapes, were particularly abundant in the Silurian. They were the
major reef builders of the late Paleozoic, finally going extinct in the early Triassic. The hexacorals, whose thecae were separated by six sepia or walls,
ranged from the Triassic to recent times and became the major reef builders
of the Mesozoic and Cenozoic eras. The tetracorals (Fig. 55), whose sepia
were arranged in groups of four, were another extinct group of reef-building
corals. The rudists (Fig. 56) were mollusks that were major reef builders in the
Cretaceous.

Tropical plant and animal communities thrived on the coral reefs.
Unfortunately, they continued to suffer from one episode of extinction after

Figure 56 *The rudists
were major reef builders
in the Cretaceous.*

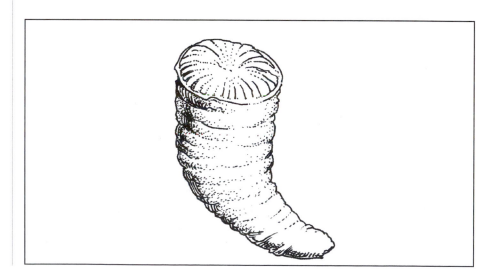

another due to their confinement to a narrow habitat range. The extinctions hit hardest organisms anchored to the ocean floor or unable to migrate out of the region. Among the extinct species were the crinoids and their blastoid relatives, which resembled tall lilies fixed to the seabed.

UPPER PALEOZOIC EXTINCTIONS

The second half of the Paleozoic followed a Silurian ice age. At this time, Gondwana wandered into the south polar region around 400 million years ago and acquired a thick sheet of ice. Glacial centers expanded in all directions. Ice sheets covered large portions of east-central South America, South Africa, India, Australia, and Antarctica. During the early part of the glaciation, the maximum glacial effects occurred in South America and South Africa. Later, the chief glacial centers switched to Australia and Antarctica.

In Australia, marine sediments were found interbedded with glacial deposits and tillites, composed of glacially deposited boulders and clay. These sediments were separated by seams of coal. This indicated that periods of glaciation were punctuated by warm, interglacial spells, when extensive forests grew. The Karroo Series in South Africa is composed of a sequence of late Paleozoic lava flows, tillites, and coal beds reaching a total thickness of 20,000 feet. Between layers of coal were fossil leaves of the extinct fern glossopteris, common on all southern continents.

The Paleozoic witnessed the advancement of the vertebrates, beginning with the first fish. Every major class of fish alive today had ancestors in the Devonian seas. However, not all species survived to the present, some becoming extinct in the intervening time. The rise of fish contributed to the decline of less mobile invertebrates, eliminating many tropical marine groups. Corals and other bottom-dwelling organisms disappeared during the middle Devonian, about 365 million years ago, possibly due to a period of climatic cooling.

The extinction appears to have extended over a period of 7 million years. Primitive corals and sponges, which were prolific limestone reef builders at the time, suffered heavily and never fully recovered. Large numbers of brachiopod families also died out when the Devonian drew to a close. The extinction did not identically affect all species that shared the same environments, however. Many Gondwanan fauna survived the onslaught due to the scarcity of reef builders and other warm-water species prone to extinction.

Several Arctic species, including certain brachiopods, starfish, and bivalves, belong to biologic orders whose roots extend hundreds of millions of years back into the Paleozoic. In contrast, tropical faunas, such as coral reef communities frequently devastated by mass extinctions, appear and disappear quite regularly on the geologic time scale.

Many extinctions correlate with glaciations. Yet no major die out occurred during the widespread Carboniferous glaciation, which enveloped the southern continents around 330 million years ago. The relatively low extinction rates were credited to a limited number of extinction-prone species following the late Devonian extinction. When the glaciers departed, the first reptiles emerged to displace the amphibians as the dominant land vertebrates. The climate of the tropics became more arid, and the swamplands began to disappear. Land once covered with great coal swamps began to dry out as the climate grew colder. The climate change set off a wave of extinctions that wiped out virtually all the lycopods.

The most profound extinction event in Earth history ended the Paleozoic 250 million years ago (Fig. 57). The loss of species was particularly hard on marine fauna. Half the families of sea life, including more than 95 percent of all known species, abruptly disappeared. On land, more than 70 percent of the vertebrates died out. So many plants succumbed to extinction that fungi briefly ruled the continents. Two distinct die-offs occurred in the space of 1

Figure 57 *The diversification of species. The large dip is in response to the great Permian extinction. Dates are in millions of years.*

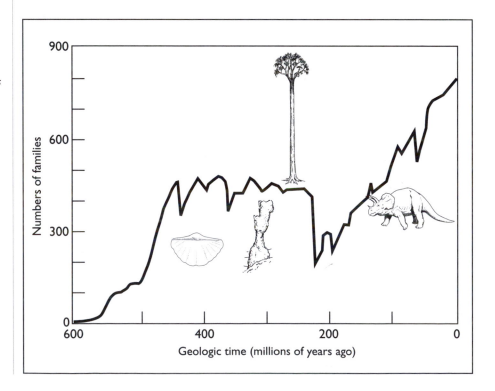

to 2 million years. About 70 percent of the species went extinct during the first event, and 80 percent of the remaining species died out in the second episode. The most pronounced loss of species took place between 252 and 251 million years ago. The final extinction pulse might have lasted less than 1 million years. The extinction left the biologic world nearly as empty of species as at the beginning of the era.

The extinction followed on the heels of a late Permian glaciation, when thick sheets of ice blanketed much of the planet, significantly lowering ocean temperatures. As a further blow, one of the largest volcanic outpourings known on Earth covered northern Siberia in thick layers of basalt. This caused considerable changes in the environment during a million-year period. The eruptions removed much of the Earth's oxygen and replaced it with choking sulfurous and carbon dioxide gases. Fossil evidence suggests that the Permian extinctions began gradually, culminating with a more rapid pulse at the end possibly due to environmental chaos.

All of the interior seas retreated from the land as an abundance of terrestrial redbeds and large deposits of gypsum and salt were laid down. Extensive mountain building raised massive chunks of crust. As sea levels fell, a continuous, narrow continental margin surrounded Pangaea, reducing the shoreline and confining marine habitats to nearshore environments. This had a major influence on the extinction of marine species. Moreover, unstable nearshore conditions resulted in an unreliable food supply. Many species unable to cope with the limited living space and food supply died out in tragically large numbers.

Groups attached to the seafloor and that filtered organic material from seawater for nutrients suffered the greatest extinction. Corals, which require warm, shallow water for survival, were hardest hit by the extinction. They were followed by brachiopods, bryozoans, echinoderms, ammonoids, foraminifers, and the last remaining trilobites. Another major group of animals that disappeared were the fusulinids (Fig. 58). These populated the shallow seas for about 80 million years. They were large, complex protozoans resembling grains of wheat and ranged from microscopic size up to 3 inches in length. Planktonic plants also went extinct. This devastated the base of the marine food web upon which other species depended for their survival.

The more mobile creatures, such as bivalves, gastropods, and crabs, escaped the extinction relatively unharmed. Those organisms, especially the shelly types, that could buffer their internal organs from changes in ocean chemistry were less likely to be wiped out and were better able to rebound after a mass extinction than their more sensitive neighbors. Nevertheless, both groups declined considerably. However, the unbuffered organisms were the hardest hit, losing 90 percent of their genera compared with 50 percent for the buffered group.

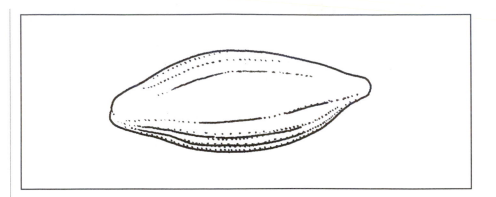

Surviving crinoids and brachiopods, which were highly prolific in the Paleozoic, were relegated to minor roles during the succeeding Mesozoic era. The spiny brachiopods, which were plentiful in the late Paleozoic seas, vanished entirely without leaving any descendants. The trilobites, which were extremely successful during most of the Paleozoic, suffered final extinction when the era ended. A variety of other crustaceans, including shrimps, crabs, crayfish, and lobsters, occupied habitats vacated by the trilobites. The sharks regained ground lost by the great extinction and continued to become the successful predators they are today.

On land, some 70 percent of the amphibian families and about 80 percent of the reptilian families also disappeared. Even the insects did not escape the carnage. Nearly one-third of their orders living in the Permian did not survive, marking the only mass extinction insects have ever undergone. The loss of plant life might have contributed to the disappearance of insects that fed on the vegetation. Following the extinction, insects shifted from a variety of dragonflylike groups, whose wings were fixed in the flight position, to forms that could fold their wings over their bodies.

MESOZOIC EXTINCTIONS

The early Mesozoic marked a rebirth of life as 450 new families of plants and animals came into existence within a relatively short span of about 1 million years. The marine world of the Triassic, from 250 to 210 million years ago, was more crowded and competitive than the wide-open oceans of the early Paleozoic. In the early Triassic, the great glaciers of the previous ice age melted, and the seas began to warm. The energetic climate facilitated the erosion of the high mountain ranges of North America and Europe. Seas retreated from the continents as they continued to rise, and widespread deserts covered the land.

In the aftermath of the Permian extinction, many once minor groups, including active predatory relatives of modern fish, squids, snails, and crabs, witnessed a rapid expansion. For example, sea urchins comparatively uncommon during the Permian are widespread in today's oceans, having sprung from a single surviving genus. Therefore, sea urchins along with other species that just managed to escape from dying out entirely would look totally different today if not for the end-Permian extinction.

Some entirely new lineages also appeared. However, rather than develop completely new body plans as organisms did during the Cambrian explosion, species merely developed new variations on already established biological forms. Although some new orders evolved, no new classes or phyla appeared. Therefore, fewer experimental organisms arose. Many lines of today's species came into existence for the first time in the geologic record.

At the beginning of the Mesozoic, the Earth was recovering from widespread glaciation. Prevailing ocean temperatures, which play a fundamental role in determining marine species diversity, remained below normal following the departure of the late Permian ice age. Invertebrates that managed to escape extinction lived in the relatively warmer equatorial seas. Corals, inhabiting the tropical waters, were particularly devastated as evidenced by the absence of coral reefs in the early Triassic.

The end-Permian extinction also decimated the ammonoids and bivalves. The ammonoids nearly became extinct again at the close of the Triassic. Of the 25 families of wide-ranging ammonoids living in the late Triassic, all but one or two families became extinct at the end of the period. The ammonoid genera that managed to escape extinction eventually evolved into scores of ammonite families, which thrived in the warm Jurassic and Cretaceous seas. Ammonite shells evolved into a wide variety of straight, helical, and coiled forms (Fig. 59) as a testament to their great diversity.

Several major groups of terrestrial vertebrates also made their debut at the beginning of the Mesozoic, including the ancestors of modern reptiles, dinosaurs, and mammals and perhaps the predecessors of birds. Once the birds became airborne, they rapidly expanded into all environments. With their superior adaptability, they successfully competed with the pterosaurs, possibly driving the flying reptiles into extinction after more than 100 million years of avian history.

During the late Triassic, around 210 million years ago, 20 percent or more families of animals tragically died out, eliminating some 50 percent of all species. Foraminifers, ammonoids, bivalves, bryozoans, corals, echinoids, and crinoids experienced global-scale extinctions. The conodonts' leechlike, jaw-shaped fossils have baffled paleontologists for centuries. They disappeared entirely after surviving the Permian extinction practically unscathed. Large

Figure 59 *A variety of fossil ammonite shells.*

(Photo by M. Gordon Jr., courtesy USGS)

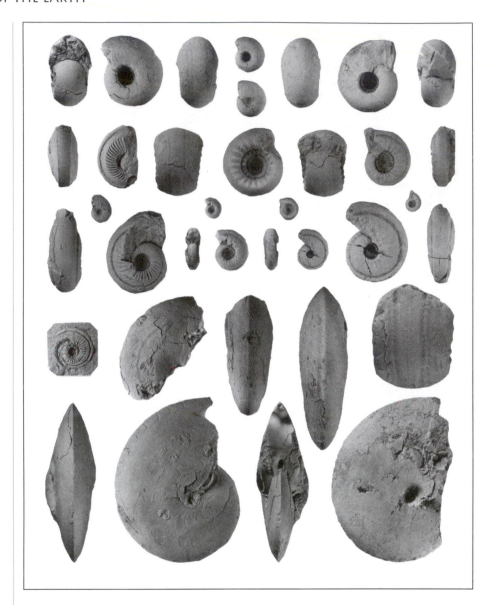

families of terrestrial animals also began dying off in record numbers, forever changing the character of life on land.

The amphibians continued to decline, with all primitive species becoming extinct early in the Triassic. The extinction occurred over a period of less than 1 million years. It was responsible for killing off nearly half the reptile families, many mammal-like reptiles, and all the thecodonts, leading the way for the ascension of the dinosaurs. The reptilelike lystrosaurus, an ancestor to mammals, was the most common land vertebrate and lived throughout Pangaea.

By the Jurassic period, which was a tumultuous time during the breakup of Pangaea, the dinosaurs were highly diversified and reached their maximum size, becoming the largest terrestrial animals ever to live. Plant cover was global, suggesting a warm, moist climate for most of the planet. The beneficial climate and high rates of vegetative growth contributed substantially to the giantism of several dinosaur species. Despite these ideal living conditions, many families of large dinosaurs, including apatosaurs, stegosaurs, and allosaurs (Fig. 60), did not survive past the end of the Jurassic.

A massive shift in the dinosaur population of North America appears to have occurred around 100 million years ago, possibly due to migrations from Eurasia. Over a period of approximately 1 million years, several dinosaurs that previously roamed the continent's enormous fern forests disappeared. They were replaced by many ancient dinosaur families that had lived in Asia and Europe and dominated the changing landscape for the next 35 million years. Gone were the giant sauropods, which were enormous, long-necked

Figure 60 *Allosauruses were among several dinosaur genera that went extinct at the end of the Jurassic.*

(Photo courtesy National Museum of Canada)

Figure 61 *Triceratops was one of the last dinosaurs to become extinct.*

dinosaurs of the Jurassic period. They were replaced by duck-billed dinosaurs, armored dinosaurs, horned dinosaurs, and huge, low-built dinosaurs with jaws designed for mowing down vegetation.

The succeeding 70 million years of the Cretaceous were the warmest of the Phanerozoic. The dinosaurs did exceptionally well during this time. However, along with many other species that thrived in the warm climate, they mysteriously vanished at the end of the period. Triceratops (Fig. 61), whose vast herds roamed the entire planet toward the end of the Cretaceous, were among the last dinosaurs to go. They might have contributed to the decline of other dinosaur species by extensive habitat destruction or by spreading diseases. Archaic mammals and birds, the dinosaurs' closest living relatives, suffered the same fate as those unfortunate beasts. Those few birds that managed to survive extinction quickly diversified into today's 9,000 avian species.

When the Cretaceous ended, the seas regressed from the land as sea levels lowered and the climate grew colder. The decreasing global temperatures and increasing seasonal variation in the weather made the world more stormy with powerful gusty winds that wrecked havoc over the Earth. These conditions might have had a major impact on the climatic and ecologic stability of the planet, possibly leading to the great extinction at the end of the era.

The extinction that brought the Cretaceous to a halt wiped out 60 to 80 percent of all living species. The die-off appears to have resulted from an environmental catastrophe such as meteorite impacts or massive volcanic eruptions that created intolerable living conditions for most land species.

Extinctions were also serious in the ocean as sea levels fell to one of their lowest levels.

After thriving for 350 million years, surviving the critical transition from Permian to Triassic, and successfully recovering from serious setbacks in the Mesozoic, the ammonites suffered final extinction at the end of the Cretaceous. At that time, the recession of the seas reduced their shallow-water habitats worldwide. The ammonites declined over a period of about 2 million years, possibly becoming extinct just before the end of the Cretaceous. They had evolved into fast-swimming, coil-shelled forms but began dying out possibly from heavy predation by the marine reptile ichthyosaur (Fig. 62). After hatching, the infant ammonites apparently remained on the surface waters among the plankton, which were heavily battered by the extinction. This great loss of plankton possibly led to the decline of the ammonites.

All shelled cephalopods were absent in the Cenozoic seas except the nautilus (Fig. 63), found exclusively in the deep waters of the Indian Ocean and the ammonite's only living relative. Shell-less cephalopod species also survived the extinction, including cuttlefish, octopuses, and squids. The squids competed directly with fish, which were little affected by the end-Cretaceous extinction.

Other major marine groups that disappeared included the rudists, which were large, coral-shaped clams that built reeflike structures. Half the bivalve genera, including clams and oysters, also died out. The gastropods, including snails and slugs, survived the extinction mostly intact. They increased in number and variety throughout the Cenozoic and are presently second only to insects in diversity.

Most warmth-loving species, especially those living in the Tethys Sea between the two major continents Laurasia and Gondwana, disappeared when the Cretaceous ended. The most temperature-sensitive Tethyan fauna

Figure 62 The ichthyosaur was an air-breathing reptile that returned to the sea.

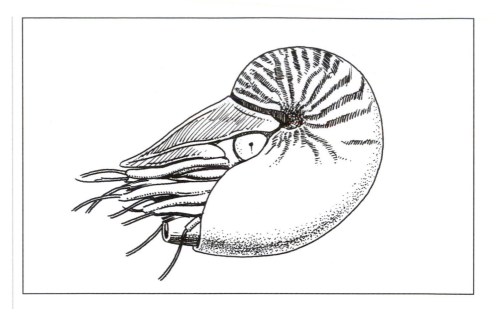

endured the heaviest extinction rates. Species that were extremely success-ful in the warm waters of the Tethys declined dramatically as ocean temper-atures plummeted.

The extinction appears to have occurred gradually over a period of 1 to 2 million years. Many animal species were already in decline, including several dinosaur and pterosaur species. Generally, no terrestrial animal weighing more than 50 pounds survived the extinction, suggesting a large body size was a severe handicap. Many plant species abruptly died out at the end of the Cre-taceous. Therefore, herbivores declined precipitously as did the carnivores that fed on them. Among the animals that survived extinction were those that ate only insects or fish.

Tropical species that rely on steady warmth and sunshine, such as coral reef communities, were especially hard hit. Although the extinction in the marine environment was severe and many animals consequently died out, few radical species evolved to take their places as happened after previous great extinctions because empty niches were simply occupied by the next of kin.

CENOZOIC EXTINCTIONS

Early in the Cenozoic, an abrupt extinction some 57 million years ago elim-inated half the unicellular, bottom-dwelling foraminifers of the deep sea,

becoming their worst calamity of the past 90 million years. In the space of ju
a few thousand years, warm and exceptionally salty water from the shallov
tropical seas flooded the deep ocean. This decimated the bottom dwellers,
which had survived the previous Cretaceous extinction 8 million years earlier
when foraminifers living on the ocean surface suffered heavily.

The world was in the midst of a gradual warming trend resulting in the
highest temperatures of the past 65 million years possibly from massive volcanic eruptions. During this time, mammals began to radiate into dazzling
arrays of new species. The small, nocturnal mammals eventually evolved into
larger animals. Some grew into giants that became evolutionary dead ends
(Fig. 64). A burst of new mammalian species in the early Eocene some 50 million years ago included many new ungulates or hoofed animals. Of the 30 or
more orders of mammals living during the early Cenozoic, only half existed
in the proceeding Cretaceous. Almost two-thirds are still living.

Mammalian evolution was not a gradual event but progressed in fits and
starts. The early Tertiary was characterized by an evolutionary lag as though
the world had not yet fully awakened after the Cretaceous extinction. By the
end of the Paleocene epoch, about 54 million years ago, mammals finally
began to diversify on a large scale. Near the end of the Eocene, about 37 million years ago, global temperatures dropped significantly. This caused a sharp
extinction that killed off many archaic mammalian species, which were large,
peculiar-looking animals that adapted poorly to the rapidly changing environment. After the extinction ran its course, the truly modern mammals began
to evolve.

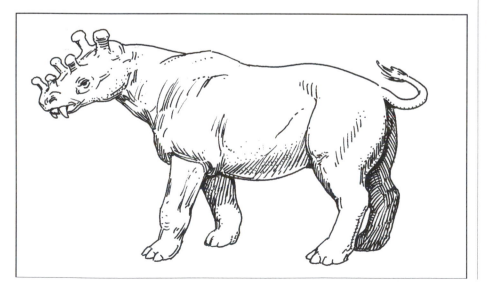

Figure 64 *An extinct
Eocene five-horned, saber-
toothed, plant-eating
mammal.*

The extinction coincided with changes in the deep-ocean circulation. It also eliminated many species of marine life on the European continent, much of which was flooded by shallow seas. The separation of Greenland from Europe might have spilled frigid Arctic waters into the North Atlantic, which significantly lowered ocean temperatures and decimated most types of foraminifers (Fig. 65). The climate grew cold. The seas withdrew from the land as the ocean dropped to perhaps its lowest level in the last several hundred million years.

Sea levels might have dropped in response to the accumulation of massive ice sheets on Antarctica, which had drifted into the south polar region and acquired a thick blanket of ice. Glaciers also grew for the first time in the highest ramparts of the Rocky Mountains. At times, Alaska connected with East Siberia at the Bering Strait, closing off the Arctic Basin from warm ocean currents and resulting in the accumulation of pack ice. A large fall in sea level caused by a major expansion of the Antarctic ice sheet led to another extinction about 11 million years ago. These cooling events removed the most vulnerable species. As a result, those species living today are highly robust, having withstood the extreme environmental swings over the last 3 million years when glaciers spanned much of the Northern Hemisphere.

The Panama Isthmus separating North and South America uplifted about 3 million years ago. It created an effective barrier to isolate Atlantic and Pacific marine species, which began to go their separate evolutionary ways. Simultaneously, extinctions impoverished the once rich fauna of the western Atlantic. Meanwhile, South America, which had been isolated from the rest of the world for 80 million years, witnessed a vigorous exchange of terrestrial species with North America across the Panama land bridge (Fig. 66). This caused many species unable to compete with the new arrivals to go extinct.

The connecting landform between North and South America halted the flow of cold-water currents from the Atlantic into the Pacific. In addition, the

Figure 65 Many types of foraminifera disappeared at the end of the Eocene epoch.

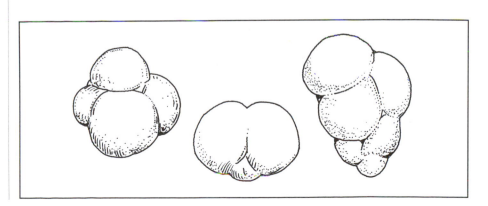

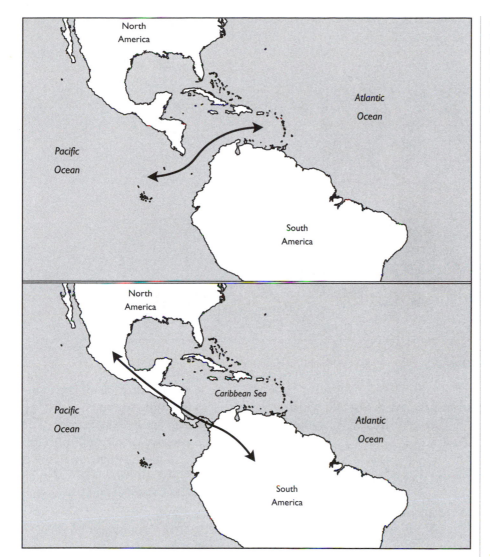

Figure 66 *The uplifting of the Panama Isthmus closed off the Atlantic from the Pacific and allowed an exchange of species between North and South America.*

closing of the Arctic Ocean from warm Pacific currents helped initiate the Pleistocene glaciation. Unlike ice ages of the past, the Pleistocene was unusual for its low extinction rates possibly because species were already well adapted to colder conditions.

As the glaciers departed at the end of the last ice age, between 12,000 and 10,000 years ago, a most unusual extinction event killed off large, terrestrial, plant-eating mammals called megaherbivores. The global environment readjusted to the changing climate with shrinking forests and expanding

Figure 67 *The extinct palorchestes was a bull-sized marsupial with extremely powerful forearms and a short trunk.*

grasslands. The climate change disrupted the food chains of many large animals. When deprived of their food resources, they simply vanished.

Woolly rhinos, mammoths, and Irish elk disappeared in Eurasia. The great buffalo, giant hartebeests, and giant horses vanished in Africa. More than 80 percent of the large marsupial mammals (Fig. 67) and a significant number of bird species vanished from Australia, which suffered the most severely of all the continents. It lost every land vertebrate species larger than a human as well as many smaller mammals, reptiles, and flightless birds.

North America lost three-quarters of all genera weighing more than 90 pounds. Gone were mastodons, woolly mammoths, 500-pound armadillos, and giant ground sloths (Fig. 68). The extinct South American ground sloth stood 12 feet tall, weighed close to 9,000 pounds, and had 7-inch deadly claws, possibly making it the largest hunting mammal ever to roam the Earth. The loss of these animals also forced into extinction their main predators: the American lion, saber-tooth tiger, and dire wolf.

As many as 135 species abruptly disappeared from North and South America in the span of just 400 years despite having lived there for millions of years. Some 35 classes of mammals and 10 classes of birds died out in North America. The extinctions occurred between 13,000 and 10,000 years ago, with the greatest die-off peaking around 11,000 years ago. Most mammals affected adversely were large herbivores weighing more than 100 pounds, many of which weighed a ton or more. Unlike earlier episodes of mass extinction, however, this event did not significantly affect small mammals, amphibians, reptiles, and marine invertebrates. Strangely, after surviving several previous ice ages over the last 2 or 3 million years, these large mammals suddenly disappeared following the last period of glaciation.

By this time, humans had become proficient hunters and roamed northward following the retreating glaciers. On their journey, they encountered an abundance of wildlife, many species of which they totally decimated. From 11,500 to 11,000 years ago, several parts of North America were occupied by Ice Age peoples, whose spear points were found among the remains of giant mammals, including mammoths, mastodons, tapirs, native horses, and camels. When people originating in Asia entered North America, they found a virgin land populated with as many as 100 million large mammals similar to those decimated in Europe and Asia. Extinctions were also massive in Australia, possibly perpetrated by the ancestors of the Aborigines.

Prior to the human invasion, Australia had been sheltered from the rest of the world for some 40 million years, drifting freely on its own. It was home to several unusual marsupials, many of which were giants in their day. What was even more striking was the extraordinary lack of large mammalian carnivores. Those roles were filled by the marsupial equivalents of wolves and lions.

Figure 68 *Giant ground sloths became extinct at the end of the Ice Age.*

By 60,000 years ago, Australia had approximately 60 species of mammals that weighed more than 20 pounds, all of which are now extinct.

The climate change resulting from the transition of Ice Age conditions to the present warm interglacial period caused sea levels to rise and the water levels to fall over much of North America. The rapid changes occurring between 15,000 and 10,000 years ago forced many species to attempt to move their ecologic ranges. Plants and animals had to respond to wide swings in climate compared with the stable climatic conditions during glaciation. Migration routes northward were no longer blocked by towering sheets of ice. Large mammals congregating at the few remaining water holes might have been vulnerable to human hunting pressures. With plentiful prey and little exposure to new diseases, human populations soared. People spread southward, leaving much big-game extinction in their wake.

After placing the extinction of species into their historical context, the next chapter will examine the various causes of mass extinctions ranging from supernovas to climate changes.

5

CAUSES OF EXTINCTION
THE FORCES OF CHANGE

This chapter examines some possible factors that influenced the mass disappearance of species. The extinction of species at specific times in Earth history appears to have celestial as well as terrestrial causes. Furthermore, the geologic time scale implies that mass extinctions are somewhat periodic, recurring every 26 to 32 million years. This extinction cycle might be related to cosmic phenomena. The gravitational attraction of an alleged sister star to the Sun could disturb a shell of comets surrounding the solar system about a light-year away and hurl them at Earth. The inevitable rain of objects onto the planet could therefore change the rules governing the evolution and extinction of life.

Terrestrial phenomena possibly explain the mass extinctions best. The geologic or rock cycle involves the circulation of convective currents in the mantle, which controls plate tectonics and consequently all activities on the surface of the Earth. The dance of the continents has been an ongoing process for 2.7 billion years and possibly longer. The breakup and assembly of continents caused dramatic environmental changes including increased volcanism that could have had a deleterious effect on climate and life.

Figure 69 *The Milky Way Galaxy.*

(Courtesy NASA)

SUPERNOVAS

As the Sun revolves around the center of the Milky Way Galaxy (Fig. 69) approximately once every 250 million years, it oscillates up and down perpendicular to the plane of the galaxy. It crosses the galactic midplane about every 32 million years, coinciding with one of the major extinction cycles. The solar system's journey through giant molecular clouds at the midplane of the galaxy might affect the Sun's output and reduce the Earth's insolation, or solar input. The limited sunlight could initiate climatic changes that dramatically affect life on the planet.

Nevertheless, the dust clouds do not seem dense enough to block out the Sun significantly during each passage through the midplane, a journey that could span several million years in duration. The solar system presently resides near the midplane of the galaxy. The Earth appears to be midway between major extinction events, the last two of which occurred approximately 37 million and 11 million years ago.

The extinction episodes might instead coincide with the approach of the solar system to its farthest extent from the galactic midplane. The extinction of the dinosaurs and large numbers of other species 65 million years ago took place at a time when the solar system's distance from the midplane was nearly maximum.

When the solar system reaches the upper or lower regions of the galaxy, it might be exposed to higher levels of cosmic radiation originating from massive exploding stars called supernovas (Fig. 70). The radiation would ionize the Earth's upper atmosphere, producing a haze that blocks out sunlight. Moreover, if a giant star such as Betelgeuse, 300 light-years away, went supernova, the Earth could receive a blast of ultraviolet radiation and X rays strong enough to burn off the ozone layer in the upper atmosphere, possibly destroying life on the planet's surface.

On a cosmic time scale, supernovas occur quite frequently in our galaxy, perhaps two or three times every century. When a giant star goes supernova, the massive nuclear explosion flings the outer layers off into space at fantastic speeds while the core compresses into an extremely dense neutron star. The supernova releases into the galaxy tremendous amounts of deadly cosmic rays, the most energetic radiation known.

The seemingly incessant rain of cosmic rays bombarding the Earth consists of atomic nuclei, protons, electrons, gamma radiation, and X rays. Most of the cosmic radiation presently striking the planet apparently originated when the giant star Vela exploded around 10,000 years ago. The supernova was 15 parsecs away, or roughly 10 times farther than the nearest star, Proxima Centauri, which is about 4 light-years from Earth.

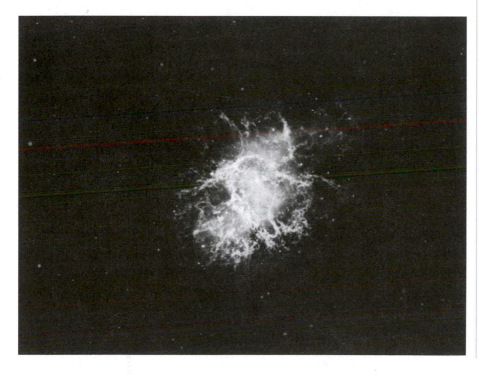

Figure 70 *The Crab Nebula in Taurus is the remains of a star that went supernova in 1054.*

(Courtesy NASA)

Around the time of the supernova toward the end of the last ice age, large mammals such as mastodons and woolly mammoths became extinct. The supernova might have emitted a burst of gamma radiation and X rays sufficiently strong to destroy upward of 80 percent of the Earth's ozone layer. The Sun's harmful ultraviolet radiation could then penetrate the atmosphere, killing off vegetation on which the large mammals depended for their survival.

GEOMAGNETIC REVERSAL

Certain geomagnetic reversals coincide with the extinction of species. A comparison between magnetic reversals with variations in the climate has shown, in many cases, a striking agreement. The Earth's geomagnetic field protects life against dangerous particle radiation from the Sun and from cosmic rays originating in outer space. The bombardment of cosmic radiation could influence the composition of the upper atmosphere by generating higher levels of nitrogen oxides, which would produce a haze that blocks out the Sun and cools the Earth.

An apparent correlation exists between changes in the magnetic field and other phenomena on the Earth's surface (Table 4). A comparison of reversals with known variations of the climate shows a striking agreement. Magnetic reversals occurring 2.0, 1.9, and 0.7 million years ago coincided with unusual cold spells as suggested by the ratio of carbon and nitrogen in ancient lake bed sediments. High ratios indicate dwindling nitrogen levels due to shrinking populations of algae and plankton in response to colder climates.

TABLE 4   COMPARISON OF MAGNETIC REVERSALS WITH OTHER PHENOMENA (DATES IN MILLIONS OF YEARS)

Magnetic Reversal	Unusual Cold	Meteorite Activity	Sea Level Drops	Mass Extinctions
0.7	0.7	0.7		
1.9	1.9	1.9		
2.0	2.0			
10				11
40			37–20	37
70			70–60	65
130			132–125	137
160			165–140	173

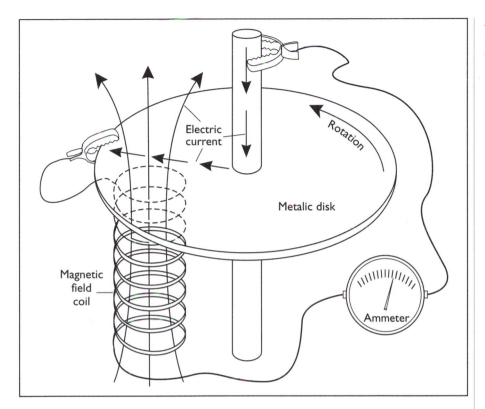

Figure 71 *The Faraday disk dynamo explains the Earth's magnetic field.*

Electric current

Rotation

Metalic disk

Magnetic field coil

Ammeter

The self-sustaining dynamo theory best describes the generation of the magnetic field in the core. A dynamo (Fig. 71) is a device that converts mechanical energy into electric or magnetic energy similar to the action of the Earth's core. The liquid core is an excellent thermal and electric conductor, which acts as a dynamo due to the Earth's rotation. Internal turmoil in the core upsets the magnetic field and causes it to reverse, possibly due to excess heat that produces erupting plumes of hot rock in the overlying mantle. During periods of intense plume activity, almost no magnetic reversals occur. However, when the plume activity is low, the number of magnetic reversals is high. Large meteorite impacts, very strong earthquakes, or intense volcanic activity have also been cited as causes of geomagnetic field reversals.

Geologic evidence from sequences of volcanic rock on the ocean floor, which record the polarity of the Earth's magnetic field when they cool and solidify, shows that the geomagnetic field has reversed often in the past. After a long, stable period of hundreds of thousands of years, the strength of the magnetic field gradually decays over a short period of several thousand years. At some point, the field collapses entirely. A short time later, it is regenerated, half the time with the opposite polarity.

Magnetic reversals have occurred throughout geologic time. Moreover, no single polarity has been dominant for long durations except possibly during the Cretaceous period between 135 and 65 million years ago, when an interval of 35 million years elapsed with no reversals. Toward the end of the Cretaceous, a long period of magnetic stability was interrupted by an abrupt reversal. During this interval, the dinosaurs were declining and eventually became extinct.

Magnetic field reversals have also been blamed for the ice ages. A reversal occurring around 2 million years ago might have initiated the Pleistocene glaciation. Reversals in the magnetic field and excursions of the magnetic poles appear to correlate with periods of rapid cooling and the extinction of species. The Gothenburg geomagnetic excursion, which occurred about 13,500 years ago in the midst of a longer period of rapid global warming toward the end of the last ice age, resulted in plummeting temperatures and advancing glaciers for 1,000 years. Apparently, this was caused by a weakened magnetic field. Presently, the Earth's magnetic field is experiencing a slow, steady decrease in intensity at such a rate that if it continues, the field could eventually collapse. A second variation in the Earth's magnetic field is a slow, westerly drift of eddies in the field, amounting to 1 degree of longitude every 5 years. The drift suggests that the fluid in the outer metallic core, which generates the geomagnetic field, is moving at a rate of about 300 feet per day.

The pattern of magnetic reversal is highly irregular and appears to result from a random process. The magnetic field reverses on average roughly two or three times every million years. Over the last 170 million years, it has reversed nearly 300 times and has done so 11 times in the last 4 million years alone. The last time the magnetic field reversed itself was about 780,000 years ago. This was around time when the huge Toba Volcano in Indonesia erupted, the greatest eruption of the past million years. The massive basalt floods of Long Valley, California and a major meteorite impact in Australasia occurred during this time. Ever since then, the Earth's climate has fluctuated between glacial and interglacial periods. Moreover, the timing of these events suggests that the Earth is well overdo for another reversal.

COMETARY IMPACTS

A shell of more than a trillion comets surrounds the Sun at a distance of about a light-year. It is called the Oort cloud, named for the Dutch astronomer Jan H. Oort, who predicted its existence in 1950, and has a combined mass of 25 Earths. The conglomeration of comets comprised of ice and rocky materials are leftovers from the creation of the solar system. They range in size up to several tens of miles in diameter and travel in highly elliptical orbits that can take them within the inner solar system (Fig. 72).

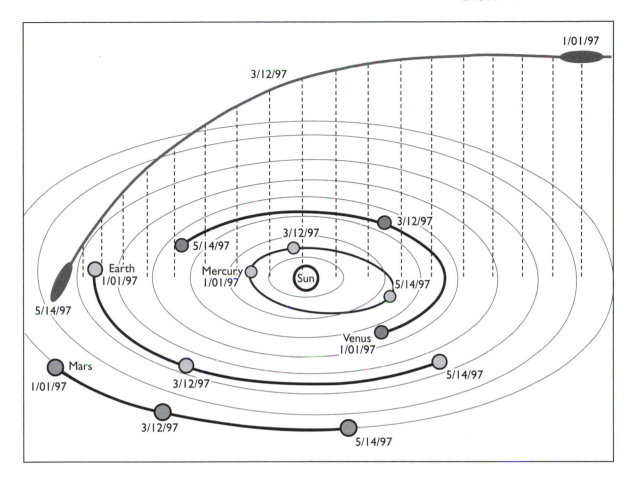

If in its orbit around the Sun the Earth is carried into the path of one or more of these icy visitors from outer space, the collision could be devastating, causing almost instantaneous extinctions. A massive comet shower, with perhaps thousands of comets impacting onto the Earth, might explain the mass disappearance of the dinosaurs and other species at the end of the Cretaceous.

As they streak toward the surface, the comets would shock heat the atmosphere, combining nitrogen, oxygen, and water vapor to form a strong nitric acid rain. The deluge of acid rain would cause a massive die out of species because most organisms cannot tolerate high acidity levels in their environment. Nitric oxide also destroys ozone. The erosion of the ozone layer by a comet shower could leave all the Earth's inhabitants exposed to the Sun's deadly ultraviolet radiation.

One explanation for the sudden mass extinctions that seem to recur roughly every 30 million years over the past 250 million years envisions a

Figure 72 Comets travel in oblique orbits with respect to the solar system.

hypothetical companion star of the Sun named Nemesis in honor of the Greek goddess who dispenses punishment over the Earth. Nemesis is thought to revolve around the Sun in a highly elliptical orbit inclined steeply to the plane of the solar system called the ecliptic. This path allows Nemesis to approach the Oort cloud about every 26 million years. The strong gravitational pull would distort cometary orbits in its vicinity, driving a swarm of comets toward the inner planets.

The apparent regularity of the cometary extinctions might also be attributed to the Earth's movement through the galactic midplane. Thick gas and dust clouds could produce gravitational anomalies strong enough to break loose comets in the Oort cloud and launch them toward Earth. The comets raining down onto the planet would wreak considerable damage, resulting in the extinction of large numbers of species.

In the depths of space far too dim to be seen by the most powerful telescopes is thought to be a tenth planetary body dubbed Planet X. It is believed to lie well outside the orbit of Pluto, perhaps 10 billion miles from the Sun. The elusive planet possibly revolves around the Sun in an elongated orbit tilted steeply to the ecliptic and takes perhaps 1,000 years to complete one revolution. Planet X is apparently no more than five times the mass of the Earth because a much larger body should have been spotted by now.

The presence of Planet X might be detected indirectly by its gravitational influence on Uranus and Neptune. These planets were deflected from their paths around the Sun during the 19th century, but no deflection was observed in the 20th century. This suggests that Planet X must be in a peculiar orbit. Perhaps every 28 million years it crosses a disk or belt of comets thought to lie in the plane of the solar system beyond the orbit of Neptune called the Kuiper belt. The gravitational attraction could disrupt the orbits of a number of these comets and send them crashing down onto Earth.

ASTEROID COLLISIONS

Asteroids are rocky remains left over from the formation of the solar system. Most lie in the main asteroid belt between the orbits of Mars and Jupiter and range in size from less than a mile to hundreds of miles wide. Asteroid impacts throughout Earth history have produced some 150 known meteorite craters scattered throughout the world (Fig. 73 and Table 5). Generally, the older the surface, the more craters it has. Major meteorite impacts also appear to be somewhat periodic, recurring every 26 to 32 million years.

How these large rock fragments managed to drop into orbits that cross our planet's path is uncertain. Apparently, for upward of a million years or more, they have revolved around the Sun in nearly circular orbits. Then, for

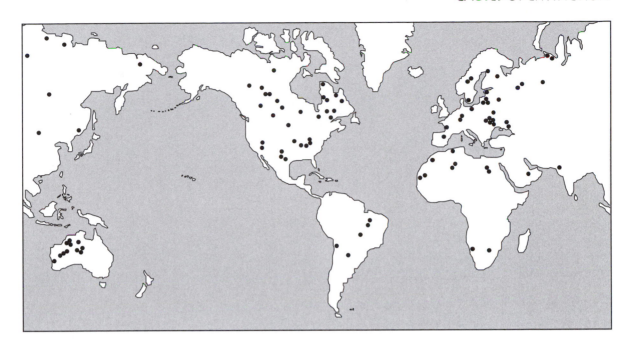

Figure 73 Location of
major impact structures.

TABLE 5   LOCATION OF MAJOR METEORITE CRATERS
OR IMPACT STRUCTURES

Name	Location	Diameter (in feet)
Al Umchaimin	Iraq	10,500
Amak	Aleutian Islands	200
Amguid	Sahara Desert	
Aouelloul	Western Sahara Desert	825
Baghdad	Iraq	650
Boxhole	Central Australia	500
Brent	Ontario, Canada	12,000
Campo del Cielo	Argentina	200
Chubb	Ungava, Canada	11,000
Crooked Creek	Missouri, USA	
Dalgaranga	Western Australia	250
Deep Bay	Saskatchewan, Canada	45,000

(continues)

TABLE 5   (CONTINUED)

Name	Location	Diameter (in feet)
Dzioua	Sahara Desert	
Duckwater	Nevada, USA	250
Flynn Creek	Tennessee, USA	10,000
Gulf of St. Lawrence	Canada	
Hagensfjord	Greenland	
Haviland	Kansas, USA	60
Henbury	Central Australia	650
Holleford	Ontario, Canada	8,000
Kaalijarv	Estonia, USSR	300
Kentland Dome	Indiana, USA	3,000
Kofels	Austria	13,000
Lake Bosumtwi	Ghana	33,000
Manicouagan Reservoir	Quebec, Canada	200,000
Merewether	Labrador, Canada	500
Meteor Crater	Arizona, USA	4,000
Montagne Noire	France	
Mount Doreen	Central Australia	2,000
Murgab	Tadjikistan, USSR	250
New Quebec	Quebec, Canada	11,000
Nordlinger Ries	Germany	82,500
Odessa	Texas, USA	500
Pretoria Saltpan	South Africa	3,000
Serpent Mound	Ohio, USA	21,000
Sierra Madera	Texas, USA	6,500
Sikhote-Alin	Sibera, USSR	100
Steinheim	Germany	8,250
Talemzane	Algeria	6,000
Tenoumer	Western Sahara Desert	6,000
Vredefort	South Africa	130,000
Wells Creek	Tennessee, USA	16,000
Wolf Creek	Western Australia	3,000

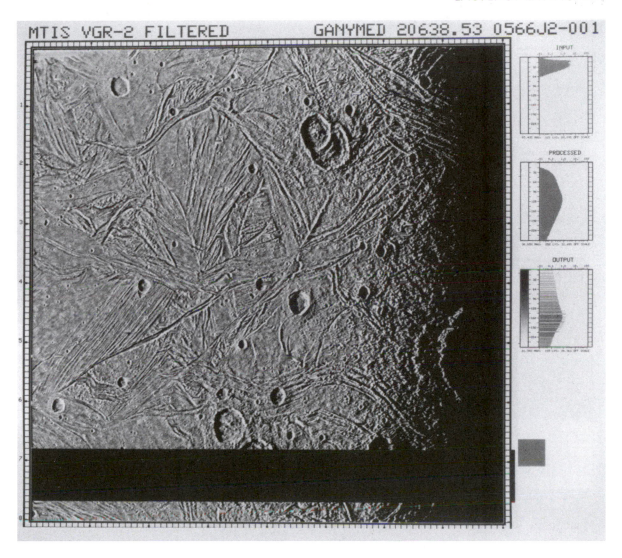

Figure 74 *A Voyager 2 digital mosaic of Saturn's moon Ganymede showing numerous craters and rills.*

(Courtesy USGS and NASA)

unknown reasons, possibly due to a passing comet or the gravitational pull of Jupiter, their orbits suddenly stretch and become so elliptical that some asteroids collide with the Earth and its moon. The craters on the Moon and other bodies in the solar system (Fig. 74) are much more apparent and numerous because they lack the weathering processes that destroy most craters on Earth and therefore remain preserved hundreds of times longer.

A large asteroid crashing down onto the Earth creates a huge explosion that ejects massive amounts of sediment and excavates a deep crater. The finer material lofts high into the atmosphere, where it shades the planet and lowers global temperatures. In addition, acids produced by the heat from a large num-

ber of meteors or comets entering the atmosphere, which generate nitrous oxides, might upset the ecologic balance by introducing strong acid rains into the environment.

About 210 million years ago, an asteroid impacting in Quebec, Canada created a crater that eventually became the 60-mile-wide Manicouagan Reservoir (Fig. 75). The impact dates to around the time when nearly half the ancient reptile families went extinct, allowing the rise of the dinosaurs. A large asteroid might have struck the Earth a second time at the end of the Cretaceous 65 million years ago near the present town of Chicxulub on the Yucatán Peninsula, Mexico. It would have created the explosive force of 100 trillion tons of TNT, or 1,000 times more powerful than the detonation of all the world's nuclear arsenals. The aftereffects of such a cataclysm would have resulted in the deaths of 70 percent of all species. Therefore, the dinosaurs might have been both created and destroyed by asteroids.

A massive bombardment of meteors or comets could also have stripped away the Earth's ozone layer in the upper atmosphere, leaving species on the surface vulnerable to the Sun's deadly ultraviolet rays. The increased radiation would kill land plants and animals along with primary producers in the surface waters of the ocean. Indeed, plankton, which are small, floating plants and animals in the sea that include dinoflagellates and foraminifers (Fig. 76), had the highest rates of extinction of any group of marine organisms, with 90 percent disappearing within half a million years following the end of the Cretaceous.

The impact of a large extraterrestrial body would result in almost instantaneous extinctions. A large asteroid striking the Earth with a force equal to a thousand eruptions of Mount St. Helens would send aloft some 500 billion tons of sediment into the atmosphere. The impact would also excavate a crater deep enough to expose the molten rocks of the mantle, creating a massive volcanic eruption. Along with huge amounts of dust generated by the impact itself, large quantities of volcanic ash injected into the atmosphere could choke off the Sun.

The compression of the atmosphere along with impact friction would provide sufficient heat to ignite global forest fires. The wildfires would consume a quarter of the land surface, turning much of the Earth into a smoldering cinder. The flames would destroy terrestrial habitats and cause extinctions of massive proportions. A heavy blanket of dust and soot would cover the entire globe and linger for months. The atmospheric pollution would cool the Earth and halt photosynthesis, killing off species in tragic numbers.

The world would have to endure a year of darkness under a thick brown smog of nitrogen oxide. Global rains would be as corrosive as battery acid. The runoff would be poisoned by trace metals leached from the soil and rock. Yet plants, surviving as seeds and roots, would be relatively unscathed. The high acidity levels in the ocean would dissolve the calcium carbonate shells of marine organisms, causing them to die out. In contrast, species with silica shells, such as diatoms (Fig. 77), would survive intact. Land animals taking refuge in burrows and creatures occupying lakes buffered against the acid would be well protected and survive the impact.

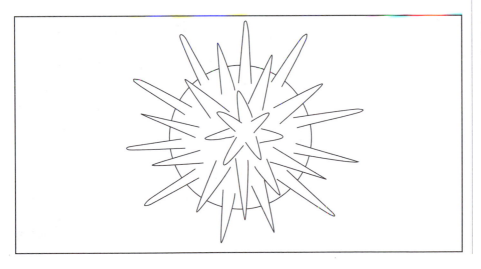

Figure 76 Plankton, including foraminifera disappeared at the end of the Cretaceous.

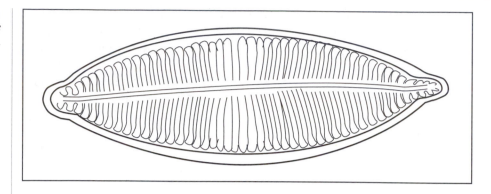

Figure 78 *The Cretaceous-Tertiary boundary rocks shown at the dark streak, just above the white sandstone in the center near Rock Springs, Wyoming.*

(Photo by R. W. Brown, courtesy USGS)

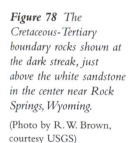

Perhaps as many as 10 or more major asteroids have struck the Earth during the last 600 million years. Moreover, the impact of a 6-mile-wide asteroid that supposedly ended the Cretaceous appears to have been a unique event. The Cretaceous-Tertiary (K-T) boundary rocks throughout the world (Fig. 78) contain a thin layer of fallout material composed of mud. Within this sediment layer are shock-impact sediments, spherules (small, glassy beads), organic carbon possibly from forest fires, a mineral called stishovite found only at impact sites, meteoritic amino acids, and an unusually high content of iridium, a rare isotope of platinum found in relative abundance on asteroids.

Whether the iridium originated from an asteroid or comet impact or from massive volcanic eruptions, which are also major sources of iridium, remains controversial. However, volcanoes do not produce the type of shock impact features on sediment grains such as those found at impact sites. The layer of spherules at the K-T boundary are up to 3 feet thick around the

alleged impact site in the Gulf of Mexico and appear to have been created by impact melt and not by volcanism.

Because stishovite, a dense form of quartz, breaks down at temperatures of about 300 degrees Celsius, far below those produced by volcanoes, it must have had an impact origin. Meteorites also contain varieties of amino acids not naturally occurring on Earth. How the meteoritic amino acids managed to escape destruction from heat generated by the impact or from ultraviolet rays after they settled onto the surface along with the rest of the impact fallout material remains unexplained.

The geologic record holds clues to other giant impacts associated with iridium anomalies that coincide with extinction episodes. However, the iridium concentrations are not nearly as strong as those at the end of the Cretaceous, which are as much as 1,000 times background levels. This suggests that the K-T event might have been unique in the history of life on Earth.

Magnetic field reversals also coincide with extinctions. Nevertheless, they do not appear to be periodic, although they might be associated with other cyclic phenomena. These include the impacts of large asteroids, which appear to account for about half the reversals. Asteroid impacts could also trigger volcanoes poised for eruption. The effects of impact volcanism might explain many characteristics of the environmental crises at important junctures in geologic time.

VOLCANIC ERUPTIONS

During the past 250 million years, 11 distinct episodes of flood basalt volcanism have occurred worldwide (Fig. 79 and Table 6). The large eruptions created a series of separate, overlapping lava flows that give many exposures a terracelike appearance called traps, from Dutch meaning "stairs." Many flood basalts lie near continental margins, where great rifts separated the present continents from Pangaea. Other massive basalt flows, such as the Columbia River basalts of the northwestern United States, which was responsible for a mass extinction 16 million years ago, are related to hot-spot activity. Hot spots are plumes of mantle rocks that rise to the surface from great depths.

The episodes of flood basalt volcanism are relatively short-lived events. Major phases last less than 3 million years. The timing of these outbursts correlates well with the mass extinctions of marine organisms. Furthermore, the volcanic episodes seem to be periodic, recurring approximately every 32 million years. During the eruption of a major basaltic lava flow, vigorous fire fountains inject large amounts of sulfur gases into the atmosphere. Moreover, flood basalts release 10 times more sulfur than explosive eruptions. The gases are converted into acid, which has severe climatic and biologic consequences.

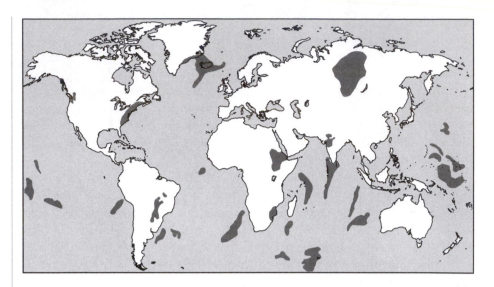

Figure 79 *Areas affected by flood basalt volcanism.*

Volcanic eruptions directly affect the Earth's climate by altering the composition of the atmosphere. Volcanoes spew massive quantities of ash and aerosols into the atmosphere (Fig. 80), which block out sunlight. Volcanic dust also absorbs solar radiation. This heats the atmosphere, causing thermal imbalances and unstable climatic conditions. Heavy clouds of volcanic dust have a

TABLE 6   FLOOD BASALT VOLCANISM AND MASS EXTINCTIONS

Volcanic Episode	Million Years Ago	Extinction Event	Million Years Ago
Columbian River, USA	17	Low-mid Miocene	14
Ethiopian	35	Upper Eocene	36
Deccan, India	65	Maastrichtian	65
		Cenomanian	91
Rajmahal, India	110	Aptian	110
Southwest Africa (Namibia)	135	Tithonian	137
Antarctica	170	Bajocian	173
South African	190	Pliensbachian	191
East North American	200	Rhaectian/Norian	211
Siberian	250	Guadalupian	249

high albedo and reflect much of the solar radiation back into space. This could shade the planet and lower global temperatures. The reduced sunlight might also cause mass extinctions of plants and animals by lowering the rate of global photosynthesis.

A reduction in solar radiation reaching the Earth's surface by as much as 5 percent could result in a drop in global temperatures of 5 to 10 degrees Celsius. Over time, this could lead to glaciation. The long-term cooling would expand the area of glaciers and lower sea levels, reducing marine habitats. The lowered temperature would also limit the geographic distribution of species and confine warmth-loving organisms to narrow regions around the Tropics.

Figure 80 *The November, 1968 eruption of Cerro Negro in west-central Nicaragua.*

(Photo courtesy of USGS)

Extensive volcanic activity 100 times more intense than the relatively mild volcanism experienced today would produce acid rain. This would cause widespread destruction of terrestrial and marine species by defoliating plants and altering the pH (acid/alkaline) balance in the ocean. Acid gases spewed into the atmosphere could also deplete the ozone layer, allowing deadly solar ultraviolet radiation to bathe the planet.

A series of huge volcanic eruptions 250 million years ago created the Siberian Traps as massive floods of lava flowed across northern Siberia within a period of less than a million years. The volcanic deposits ranging from 250 feet to 2 miles thick include some 45 separate flows that cover at least 350,000 cubic miles. It was the largest known eruption on land over the past half billion years and disgorged more lava than a million Mount St. Helens volcanoes.

The volcanic outburst also coincided with the great extinction at the end of the Permian period, when 95 percent of all species vanished. The eruptions might have blanked the Earth with sun-blocking sulfate gases and dowsed it with deadly acid rain. Furthermore, ocean carbon dioxide concentrations apparently reached 30 times present-day levels, causing deaths among marine species unable to handle the carbon dioxide poisoning. The hardest hit were animals with passive respiration such as corals. In contrast, active breathers like snails and clams generally escaped extinction.

At the end of the Triassic period, around 210 million years ago, another volcanic catastrophe sent massive rivers of lava oozing out of giant fissures in the ground and paved over a continent-sized swath of land. This eruptive outpouring was one of the few great basalt flows in Earth history. During this time, all continents were locked in a single landmass called Pangaea, which began to tear apart at the seams. Within a short span of a few million years, basalt erupted along the central spine of the supercontinent, spreading over an area of about 2.7 million square miles or nearly the size of Australia.

The giant floods of lava could have released so much carbon dioxide into the atmosphere that the climate altered dramatically. Indeed, the fossil record suggests that carbon dioxide concentrations surged at the end of the Triassic. The eruptions also coincided with one of the largest known mass extinctions, when more than half the species on Earth disappeared, including the carnivorous reptiles. This led to the dominance of the meat-eating dinosaurs.

Another major flood basalt eruption created the gigantic Ontong Java Plateau under the Pacific Ocean in Southeast Asia. It occurred in two phases: about 120 million years ago and again about 90 million years ago, the latter of which coincides with a major extinction. Massive volcanic eruptions have also been cited as the principal cause for the dinosaur extinction. At the end of the Cretaceous, a giant rift ripped open the west side of India. Huge volumes of molten lava poured onto the surface. Nearly 400,000 square miles of lava were

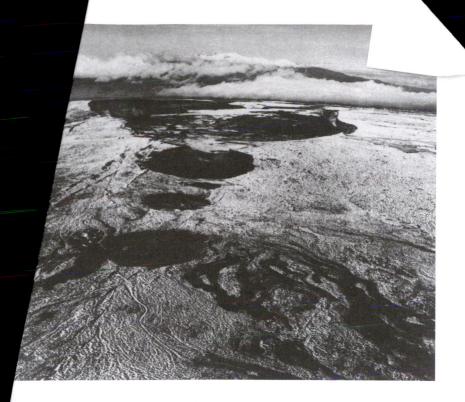

...tures, ocean currents, productivity, and many other factors of fundamen-
...importance to life.

Continental motions had a major influence on the distribution, isola-
...on, and evolution of species. Many different environments result in a wide
...ariety of species. The continual variations in world ecology had a substantial
...nfluence on the course of evolution and accordingly on the diversity of living organisms. Therefore, evolutionary trends varied throughout geologic time
in response to major environmental changes. Natural selection acted to adapt
organisms to the new conditions forced onto them by environmental factors
that were affected, in large part, by continental drift.

Continents and ocean basins are continuously being reshaped and
rearranged by crustal plates in motion. The changing shapes of the ocean
basins by the movement of continents also influences the flow of ocean currents (Fig. 82), the width of continental margins, and the abundance of marine
habitats. When continents rift apart, they override ocean basins, making seas
less confined. This raises global sea levels and lowers the land. The rising seas
inundate low-lying areas inland of the continents, dramatically increasing the
shoreline and shallow-water marine habitat area. The expansion of the inhab-
itable area can thus support a larger number of species.

ejected over a period of ~
India known as th

Simu'
separate fro
along the co
ously. Evidenc
region from the
of the globe mig
bility of the plane

A sharp drop
percent down to 29
canic activity and se
extinction of the dino:
A significant change in t
during volcanic eruptic
Atlantic, when many de
extinct. At this time, ocean
during the past 70 million ye

The shocked quartz ar.
could have originated from ma
dreds of thousands of years. As
canic eruptions can produce sho
generated by volcanic eruptions an
large asteroid impacts. Therefore, c
tions etched across shocked quartz g
created by impacts.

Volcanoes whose magma source
Mauna Loa on Hawaii (Fig. 81), produce
osmium, another rare element found in th
pherules at the K–T contact believed to hav
also have been produced by volcanic eruptio.
have been killed off without the aid of an ast.

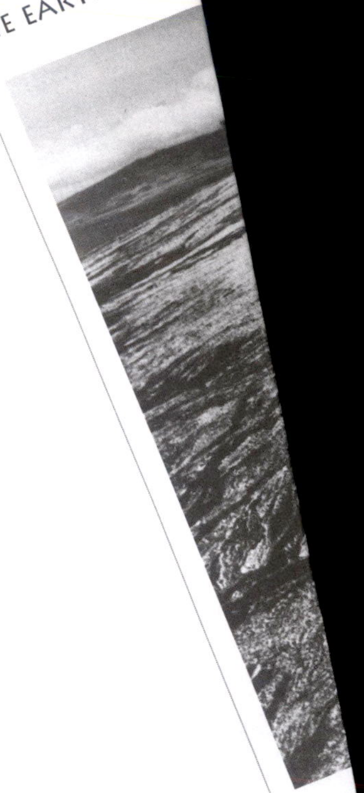

Figure 81 The Mauna
Loa Volcano, Hawaii.
(Courtesy USGS)

per
tal

ti
v

CONTINENTAL DRIFT

Perhaps the greatest force that has influenced evoluti
ing of the continents. Plate tectonics and continental
nent roles in the history of life practically since the ve
in the relative configuration of the continents and the c
ing influence on the environment, climate conditions, an
species. The changes in continental shapes significantly a

The assembly of landmasses into a supercontinent permits the free flow of ocean currents. These distribute heat from the tropics to the poles, maintaining more uniform global temperatures. The ocean basins also widen, causing sea levels to drop. A substantial and rapid fall in sea levels could have both a direct and indirect influence on the biologic world. It would increase seasonal extremes of temperature on the continents, thereby elevating environmental stress on terrestrial species. A falling sea level and the retreat of inland seas result in a continuous, narrow continental margin around the supercontinent. This, in turn, reduces the shoreline, radically limiting the marine habitat area. In addition, unstable nearshore conditions make food supplies unreliable.

Figure 82 *The major ocean currents are controlled by the positioning of the continents.*

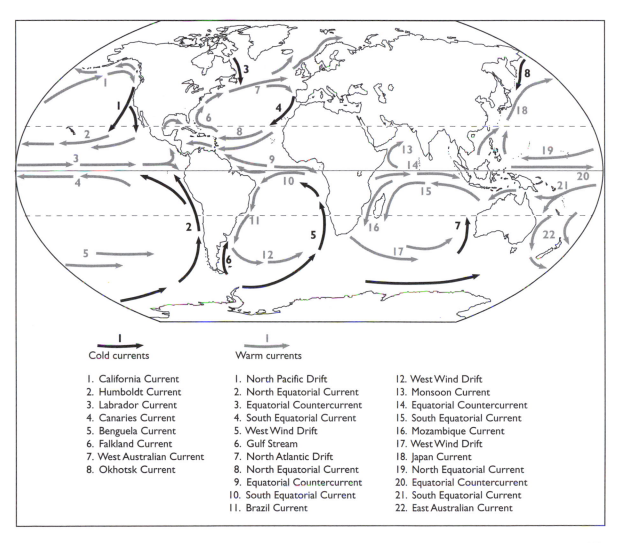

Cold currents   Warm currents

Cold currents	Warm currents	
1. California Current	1. North Pacific Drift	12. West Wind Drift
2. Humboldt Current	2. North Equatorial Current	13. Monsoon Current
3. Labrador Current	3. Equatorial Countercurrent	14. Equatorial Countercurrent
4. Canaries Current	4. South Equatorial Current	15. South Equatorial Current
5. Benguela Current	5. West Wind Drift	16. Mozambique Current
6. Falkland Current	6. Gulf Stream	17. West Wind Drift
7. West Australian Current	7. North Atlantic Drift	18. Japan Current
8. Okhotsk Current	8. North Equatorial Current	19. North Equatorial Current
	9. Equatorial Countercurrent	20. Equatorial Countercurrent
	10. South Equatorial Current	21. South Equatorial Current
	11. Brazil Current	22. East Australian Current

Mountain building associated with the movement of crustal plates alters the patterns of river drainages and climate, which, in turn, affect terrestrial habitats. Raising land to higher elevations, where the air is thin and cold, can initiate glacial activity, especially in the higher latitudes. During times of highly active continental movements, volcanic activity increases. This volcanism occurs especially at midocean rifts, where tectonic plates pull apart by upwelling magma from the upper mantle. Extensive volcanism affects the composition of the atmosphere, the rate of mountain building, and the climate. The positioning of the continents with respect to each other and to the equator determines climatic conditions. When most of the land gathers around the equator, the climate is warm and hospitable to life. Conversely, when lands wander into the polar regions, the climate turns cold, spawning episodes of glaciation and mass extinction.

CLIMATE COOLING

Among the most important factors influencing the diversity of species is global temperature. Lower seawater temperatures limit the geographic distribution of marine species. An episode of climatic cooling could extinguish species not adaptable to the new, colder conditions or unable to migrate to warmer refuges. Those that manage to escape extinction are restricted to narrow margins near the equator.

Certain extinction events coincide with episodes of glaciation (Table 7). The effects of global cooling on life are considerable. The living space of warmth-loving species is limited to areas narrowly confined to the tropics. The loss of habitat area and the number of species it supports explains why extinctions generally follow periods of climatic cooling.

Species that fail to adapt to colder conditions are usually the hardest hit. Because lowered temperatures slow the rate of chemical reactions, biologic activity during a major glacial event is expected to function at a lower energy state, which would affect species diversity. The climate change would also spur the development of new species better adapted to the colder conditions.

The extinction of the dinosaurs at the end of the Cretaceous was apparently caused by a radical change in climate, which killed off plants that were the dinosaurs' main food source. Mammals, because they supplied their own body heat, were particularly well suited to a colder environment and rapidly filled the available space vacated by the dinosaurs. Other animals such as birds and aquatic species developed migratory habits, enabling them to escape the frigid weather seasonally for warmer climes elsewhere.

As the world's oceans cool, mobile species tend to migrate into the warmer regions of the tropics. Species unable to travel or that are trapped in

TABLE 7   THE MAJOR ICE AGES

Time in Years	Event
10,000–present	Present interglacial
15,000–10,000	Melting of ice sheets
20,000–18,000	Last glacial maximum
100,000	Most recent glacial episode
1 million	First major interglacial
2 million	First glacial episode in Northern Hemisphere
4 million	Ice covers Greenland and the Arctic Ocean
15 million	Second major glacial episode in Antarctica
30 million	First major glacial episode in Antarctica
65 million	Climate deteriorates, poles become much colder
250–65 million	Interval of warm and relatively uniform climate
250 million	The great Permian ice age
700 million	The great Precambrian ice age
2.4 billion	First major ice age

enclosed basins are generally the hardest hit by extinction. Those species that have previously adapted to cold conditions still thrive in today's chilly oceans. Most are herbivores that tend to be generalized feeders that consume many types of vegetation.

Not all climatic cooling resulted in glaciation, which is determined by such factors as the positioning of the continents, the tilt of the rotational axis, and the ellipticity of the Earth's orbit. Nor did all extinctions follow lowering sea levels caused by growing glaciers. At the beginning of the Oligocene epoch, beginning about 37 million years ago, seas that had invaded the continents drained as the ocean withdrew to one of its lowest levels during the past several hundred million years due to the mass accumulation of ice atop Antarctica.

Although sea levels remained depressed for 5 million years, almost no excess extinction of marine life occurred. Therefore, crowding conditions brought on by lowering seas cannot be responsible for all extinctions. Furthermore, during many mass extinctions, the sea level was not much lower than it is today.

During the final stages of the Cretaceous, when the seas were departing from the land and the level of the ocean began to drop precipitously, temperatures in a broad tropical ocean belt called the Tethys Sea began to fall. The cooler temperatures might explain why the Tethyan species that were the most temperature sensitive suffered the heaviest extinction rates. Many species that were amazingly successful in the warm waters of the Tethys were totally decimated as temperatures dropped. Following the extinction, marine species assumed a more modern appearance as ocean bottom temperatures continued to plummet.

After listing some possible causes of mass extinctions, the next chapter will examine what effects they have on life, including species survival, radiation of species, and influence on the fossil record.

6

EFFECTS OF EXTINCTION
THE INFLUENCE ON EVOLUTION

This chapter examines how mass extinctions influenced the outcome of life on Earth. Extinction is an inevitable part of evolution and essential for the advancement of species. Therefore, little happens in evolution without extinction first disrupting ecosystems and a large number of species that occupy them. Each mass extinction marked a watershed in the evolution of life. Whole groups of species fell by the wayside, giving survivors a chance to rule the world until they met their demise.

Practically every species that ever existed has gone extinct, setting the stage for the evolution of entirely new life forms. Since the first appearance of life, extinctions have eliminated organisms, prompting the development of species better adapted to exploit their environment to its fullest. Consequently, extinctions have played an enormous role in the evolution of life. Therefore, the more devastating and globally encompassing an extinction event, the greater the evolutionary change.

All extinction events appear to indicate biologic systems in extreme stress brought on by a radical change in the environment. After a major die-off, new species evolve to replace those that failed. Had species not become

Figure 83 *Marine life at the 100-foot depth at the Point Loma kelp beds off San Diego, California.*

(Photo by R. Outwater, courtesy U.S. Navy)

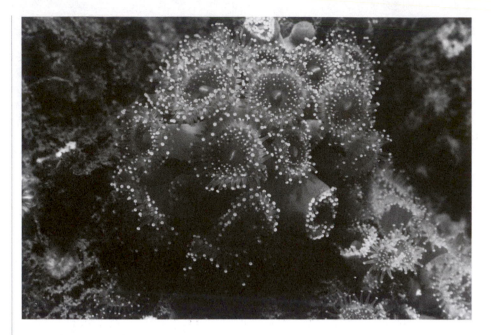

extinct to make room for more advanced organisms, life would not have progressed to the myriad of biological forms in existence today (Fig. 83). The only organisms would be simple microscopic creatures in the sea, and life would be much the same as when it began.

EXTINCTION EPISODES

Throughout Earth history, species have regularly come and gone on geologic time scales. Those living today represent only a small fraction of all species of plants and animals thought to have existed in the geologic past. Each mass extinction resets the evolutionary clock, forcing species to start anew. When a major extinction event occurs, new species develop to fill vacated habitats. Because of their significant impacts on life, major extinction events also mark the boundaries between geologic periods.

Most mass extinctions followed periods of environmental upheavals such as global cooling, even though a cooler climate was not necessarily accompanied by glaciation. During periods of glaciation, sea levels fell significantly because large quantities of the ocean's water were locked up in the expansive ice sheets. The lowered level of the ocean reduced the area of shallow-water habitats, forcing crowding conditions and dwindling nutrients worldwide. Yet sea levels were no lower than they are today during many extinctions. The drop in temperatures also restricted the geographic distribution of species, confining the most temperature sensitive to warmer regions bordering the equator.

In the late Precambrian about 670 million years ago, when animal life was still sparse, the first mass extinction decimated the ocean's population of singled-celled phytoplankton, which were the first organisms with cells containing nuclei. The mass disappearance of these species coincided with extensive glaciation. When the glaciers departed near the end of the era, a mass diversity of organisms gave rise to a large variety of new species whose unique characteristics have never been seen during any subsequent time in Earth history.

The vast majority of the Earth's fauna and flora lived during the Phanerozoic eon from about 570 million years ago to the present. This was a period of phenomenal growth as well as tragic episodes of mass extinction. Each involved the loss of more than half the species living at the time. Five major extinctions, interspersed with five or more minor die-offs, occurred during this interval.

The first mass extinction, around 530 million years ago, decimated more than 80 percent of all marine animal genera. This became one of the worst extinctions in geologic history. A second mass dying at the end of the Ordovician period about 440 million years ago eliminated some 100 families of marine animals. Another major die-off during the middle Devonian period about 365 million years ago witnessed the mass disappearance of many tropical marine groups.

The greatest loss of life in the fossil record killed off half the families comprising more than 95 percent of all marine and 80 percent of terrestrial species at the end of the Permian period about 250 million years ago. Another tragic event at the end of the Triassic period about 210 million years ago took the lives of nearly half the reptilian species. The most familiar extinction eliminated the dinosaurs along with 70 percent of all known species at the end of the Cretaceous period about 65 million years ago. A major disruption cut down the mammals about 37 million years ago and again as recently as 11 million years ago.

PERIODIC EXTINCTIONS

The fossil record suggests that mass extinctions might be periodic. They could be caused by celestial influences such as cosmic rays from supernovas or by major meteorite or comet impacts. Ten or more large asteroids or comets have collided with the Earth over the last 600 million years. Analysis of 13 major impact craters distributed over a period from 250 million to

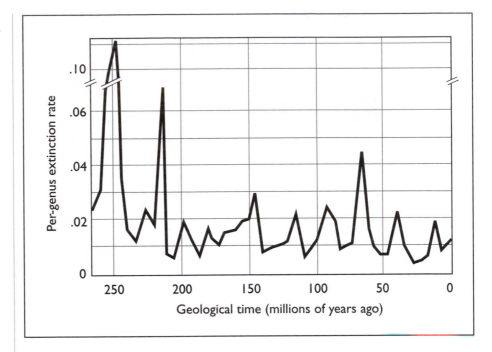

5 million years ago suggests a cratering rate of roughly one every 28 million years.

Eight significant extinction events have occurred since the great Permian catastrophe. Many of the strongest peaks coincided with the boundaries between geologic periods. The episodes of extinction appear to be cyclical, occurring every 26 to 32 million years (Fig. 84). Longer intervals of 80 to 90 million years are related to the breakup and collisions of continents. Exceptionally strong extinctions occur every 225 to 275 million years, corresponding to the solar system's orbital period around the center of the Milky Way Galaxy.

The extinctions, however, might simply be episodic with relatively long periods of stability followed by random, short-lived extinction events that only seem periodic. Major extinctions could therefore reflect a clustering of several minor events at certain times. Due to the nature of the fossil record, these might only mimic a cyclical pattern. In other words, random groupings of extinct species on a geologic time scale that is itself uncertain could merely be coincidental. Furthermore, a short period of rapid evolution might manifest itself in the geologic record as though preceded by a mass extinction when indeed none had occurred.

When the fossil record suggests the demise of large numbers of species at roughly the same time, paleontologists often invoke some sort of catastrophe as the cause of extinction. Many geologists are beginning to accept catastrophes as normal occurrences in Earth history and as a part of the uniformitarian process, also called gradualism. Certain periods of mass extinctions appear to be the result of some catastrophic event, such as the bombardment of one or more large asteroids or comets, rather than due to subtle changes, such as a change in climate or sea level or an increase in predation.

Paleontologists search for environmental factors to explain the sudden emergence of a variety of new species following extinction. However, some large changes in an ecosystem do not always occur because of a cataclysm but because a minor event initiated a chain reaction that rapidly leads to widespread disruption. This occurs because the evolution of a single species also affects the evolution of species with which it interacts. A given species might remain unchanged for long periods and then, due to a brief spurt of mutations, experience a period of rapid change that affects other species around it. Therefore, large events in evolutionary history might simply reflect the natural fluctuations in an ecosystem rather than the consequences of catastrophes.

BACKGROUND EXTINCTIONS

Since life's first appearance, a gradual loss of species known as background or normal extinction has been a common feature of living beings. Periods of lower extinction rates are punctuated by major extinction events, and the difference between the two is only a matter of degree. Species regularly come and go even during optimal conditions. Those that became extinct might have lost their competitive edge and were replaced by a superior, better adaptable species.

A qualitative as well as a quantitative distinction defines the difference between background and major extinctions. Mass extinctions are not simply intensifications of processes operating during background times, however. Survival traits developed during periods of lower extinction rates become irrelevant when mass extinctions occur. Mass extinctions therefore appear to be less discriminating of species with respect to the environment than are normal extinctions. Different processes might be operating during times of mass extinction than during normal extinctions. Moreover, the same types of species that succumb to mass extinctions also succumb to background extinctions—only a lot more of them.

Species that suffer extinction might have been developing certain unfavorable traits during background times. Even within the same species, daughter species might develop superior survival skills and replace their parent species.

Figure 85 *Fossil brachiopods and trilobites from the Bonanza King Formation, Trail Canyon, Death Valley National Monument, Inyo County, California.*

(Photo by C. B. Hunt, courtesy USGS)

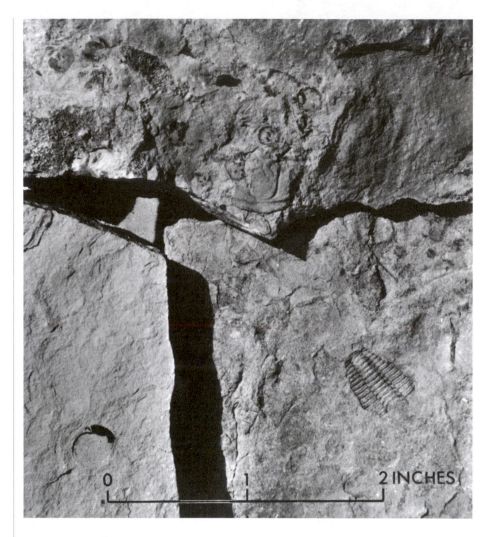

The distinction between background and mass extinctions could also be distorted by ambiguities in the fossil record, especially when certain species are favored over others for fossilization. Only under demanding geologic conditions that promote rapid burial with little predation or decomposition are the bodies of dead organisms preserved to withstand the rigors of time (Fig. 85). Because species with hard body parts fossilize better than soft-bodied organisms, they are more favorably represented in the fossil record and therefore present a skewed account of historical geology. Furthermore, the fossil record can be extremely insensitive to major changes in groups of organisms. For example, a family containing some 60 species could be nearly completely devastated, leaving only one surviving species, and yet no change would be recorded in the fossil record.

Episodes of extinction appear to be virtually instantaneous in the fossil record because a resolution of several thousand years spanning millions of years of geologic time is not possible. More likely, the extinctions occurred over lengthy periods of perhaps a million years or more. Due to erosion or nondeposition of the sedimentary strata that preserve species as fossils, the die-offs only appear sudden. Several times in Earth history, falling sea levels reduced sedimentation rates and the preservation of species. Therefore, a sudden break in geologic time might, in reality, have extended over a lengthy period.

Following a mass extinction, surviving species radiate outward to fill vacated habitats, spawning the development of entirely novel organisms. These new species might develop special adaptations that give them a survival advantage over others. These adaptations could lead to exotic species that prosper during intervals of normal background extinctions. However, because of over-specialization, they are not adequately equipped to survive mass extinction. Accordingly, the fossil record is replete with many unusual creatures (Fig. 86).

EXTINCTION SURVIVAL

Most extinction episodes appear to select certain categories of species for execution. The analysis of victims versus survivors might ultimately lead to the

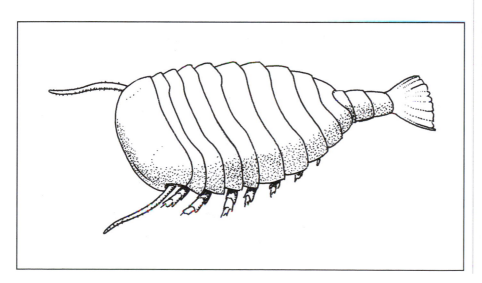

Figure 86 Sidneyia was an unusual arthropod and predominant predator of the Burgess Shale Fauna.

Figure 87 *Worldwide distribution of coral reefs, which contain diverse groups of related species.*

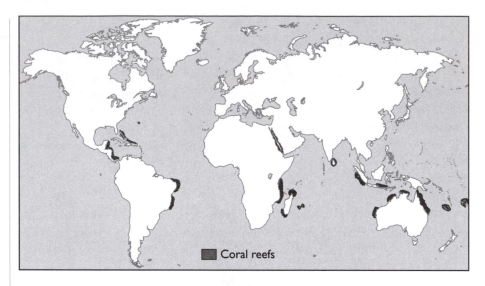

Figure 87 *Worldwide distribution of coral reefs, which contain diverse groups of related species.*

Coral reefs

primary causes of the die-off. Surviving species are particularly hardy and more resilient toward subsequent environmental changes. They tend to occupy large geographic ranges that contain many groups of related organisms (Fig. 87). However, just because a species escaped extinction does not necessarily mean it possessed superior survival traits or was better suited to its environment. Instead, the losers might have been developing certain unfavorable characteristics that made them more vulnerable to extinction.

Extinctions not only reduce the number of different species but also the total number of species. Once an extinction event occurs, biological systems seem temporarily immune to further random cataclysms. Moreover, after a major extinction, fewer species remain to die out. Therefore, until the evolution of large numbers of species, including extinction-prone types, any intervening catastrophes would have comparatively little effect. After each extinction, the biological world requires a recovery period before it is again ready to face another extinction.

Following the mass extinction that ended the Paleozoic, leaving the world with nearly as little species diversity as when the era began, life witnessed many remarkable advancements. Surviving species possessed similar attributes of contemporary populations. Many were the same types that survived the end-Cretaceous extinction, suggesting they might have perfected survival characteristics lacking in those species that went extinct. Species living in the sea were especially insulated and unaffected by extreme environmental changes due to the ocean's large thermal momentum, allowing it to retain heat for long periods.

As the world recovered from the late Permian extinction, several regions of the ocean began to fill with many specialized organisms. The overall diversity of species rose to unprecedented heights. However, rather than develop entirely novel forms as species did during the Cambrian explosion, when nearly all present phyla came into existence, those that survived the Permian extinction acquired morphologies or body plans based on simple skeletal types. This resulted in fewer experimental organisms and therefore less extinction-prone types.

Although the dinosaurs suffered final extinction at the end of the Cretaceous, they might not have done anything "wrong" biologically. The dinosaurs were keenly adapted to their environment, which accounted for their great success and enabled them to dominate the world for 170 million years. Had the dinosaurs escaped extinction, a certain small carnivorous dinosaur called stenonychosaurus (Fig. 88), with ratio of brain mass to body weight characteristic of early mammals, could conceivably have suppressed mammalian evolution. Worse yet, the continued dominance of the dinosaurs would ensure that our own species would never have come into existence.

A sudden change in environmental factors might have dealt the dinosaurs a final blow because of their inability to adapt to the new conditions rapidly. Many dinosaur species were apparently already in decline millions of years prior to the extinction that finally finished them off. In effect, they were biologically well suited for the conditions of the Mesozoic era but were incapable of adapting to the environment of the Cenozoic.

Figure 88
Stenonychosaurus had a large ratio of brain weight to body weight.

125

Marine species that evolved into better-adaptive forms survived the Cretaceous extinction and possessed characteristics similar to those living in the Mesozoic seas. Although the extinction in the oceans was severe and many species died out accordingly, few radical forms appeared simply because closely related species occupied the vacated habitats. Species that inhabited unstable environments, especially in the higher latitudes, were extremely successful. In contrast, those living in tropical coral communities (Fig. 89) were ceaselessly battered by extinction.

The instability of the tropical environment appears to be responsible for creating species at a faster pace than other regions. Highly disturbed, shallow-water environments might also breed species more resistant to extinction. Shallow, nearshore waters, where fluctuations in sea level, severe storms, and climate change cause frequent and dramatic shifts in habitat, are responsible for the origination of more orders of animals than deep, offshore waters, where environmental changes are subdued.

The higher the resistance to extinction the fauna of a particular region have, the better their chances of developing novel forms that survive to give rise to new species. Therefore, the tropical reef environment is the likely cradle of marine species diversity even though it is usually hardest hit by extinction. This is because as diversity increases, speciation declines and extinction rates rise. Species are better off when diversity is low, in part, because they face greater competition from other species.

Surviving species require a recovery period before they are ready to face another mass extinction. This is why the mammals took so long to finally diversify after the dinosaurs died out. After the extinction, the number of mammalian genera rapidly expanded to a high of about 130 genera some 55 million years ago, during a time when the Earth experienced an unusual warming spell. Unfortunately, rapid diversification, once the mammals began to occupy niches vacated by the dinosaurs, led to overspecialization in many cases. As a result, some mammals lacked the proper survival traits to withstand the next extinction event around 37 million years ago.

Each mass extinction restarts the evolutionary cycle, forcing life to start anew. Surviving species radiate outward and fill empty habitats, which subsequently lead to entirely new varieties. Genetics rather than morphology defines a species. Different organisms might share similar physical attributes only because they occupy the same environment. For example, when reptiles and mammals returned to the sea, they assumed the appearance of fish (Fig. 90).

Furthermore, the amount of genetic information each cell contains steadily rises from simple to complex organisms. For instance, a radical jump in the number of genes between the invertebrates and vertebrates some 500 million years ago doubled the gene count. All invertebrates have fewer genes

...ans, possess a similar

...ng process through–
...disappearing in large
..., extinctions play a
...jor extinction event
...ll vacancies left by

...ss than 1 percent of
...ies believed to have
...n, a period of robust

Figure 90 When mammals took to a life in the sea, they assumed the appearance of fish. Shown here are the sperm whale (top) and the blue whale (bottom).

than vertebrates. However, all vertebrates, including hun number of genes.

RADIATION OF SPECIES

The emergence and radiation of species has been an ongo out Earth history. Species have regularly come and gone, numbers at certain junctures in geologic time. Therefor fundamental role in the evolution of species. When a ma occurs, new species develop and radiate outward to f extinct organisms.

Plants and animals living on Earth today represent l all species that have ever existed. Most of the 4 billion spe lived in the geologic past arose during the Phanerozoic eo evolution as well as excruciating extinctions. During the Cenozoic, species diversity reached unprecedented heights, filling our world with an immense variety of fauna and flora.

Between 5 million and 30 million species are believed to inhabit the Earth today. Most conduct their lives completely out of view. Many play critical roles in food chains and make vital nutrients available to higher organisms. These simple creatures, including bacteria, fungi, and plankton, comprise about 80 percent of the biomass, or the total weight of all living matter. Furthermore, marine phytoplankton (microscopic unicellular plants) produce most of the breathable oxygen available on the planet.

The geologic record implies that new forms of life are constantly emerging. When one species fails, it goes extinct—never to be seen again. Once a species becomes extinct, it is forever lost because the chances of its unique combination of genes reappearing are astronomically low. Even if the environment in the future ideally matched the warm conditions of the Mesozoic when the dinosaurs were prolific, in all likelihood, dinosaurs will never return. This is instructive for us today. If we cause too many species to go extinct, the loss of diversity could match that at the end of the Cretaceous.

Thus, the evolutionary process seems to progress in the direction of greater complexity. Although evolution perfects species to live at their optimum in their respective environments, it cannot return to the past. However, convergent evolution does enable one species to resemble an entirely different species physically only because they share the same environment. For example, when the reptiles and mammals reentered the sea, they acquired the attributes of fish.

The dinosaurs and mammals coexisted for some 160 million years. The dinosaurs were not the only creatures to go extinct, however. A large percentage of other species vanished simultaneously along with them. Therefore, something in the environment made them all unfit to survive, yet that factor did not significantly affect the mammals and other small animals. Birds went almost completely extinct along with the dinosaurs, their closest relatives. Few fossils of modern birds predate the mass extinction. Only a couple dozen or so bird lineages came through the great die-off, most of which had their beginnings just millions of years earlier.

Generally, the smaller and more widespread the species, the better its chances of survival. When the dinosaurs died out, the mammals, due to their higher evolutionary advancement, their tiny size, and their larger numbers, sailed through the mass extinction practically unscathed. At least 100 terrestrial vertebrate lineages survived the K–T boundary, including rodents, primates, and other mammals. By the luck of the draw, the mammals happened to be at the right place at the right time to take over the world.

During the reign of the dinosaurs, most mammals were small, nocturnal creatures with a limited range. However, they possessed certain survival characteristics that allowed them to replace the dinosaurs without themselves succumbing to extinction. One of these traits might have been their higher intelligence, although many dinosaur species appeared to have been suitably intelligent as well. The mammals were also warm-blooded, an important survival characteristic, and so were apparently some dinosaur species.

These adaptations appear to have given the mammals a decisive advantage during times of environmental stress. For some reason, though, similar characteristics were ineffective for dinosaur survival. Perhaps the small size of the mammals helped them survive the extinction. Yet, not all dinosaurs were giants, and many were no larger than most mammals living today. The mammals might have preyed on dinosaur eggs, thereby winning the battle of evolution by attrition. Whatever were the mammals' advantages, they apparently outcompeted and outpopulated the dinosaurs and, perhaps over an extended period of several million years, completely replaced them.

The placental mammals appeared to survive the end-Cretaceous extinction and flourish at the expense of the more primitive marsupial mammals, which were decimated by the catastrophe (Fig. 91). Supposedly, the more advanced placental reproduction, with the embryo developing in utero, was a major contribution to placental mammals' great success. Yet some dinosaurs species are thought to have given live births as well. Moreover, egg-laying fish, amphibians, reptiles, and birds still enjoy a high degree of reproductive success as indicated by the large populations of these animals throughout the world today.

Figure 91 *Primitive marsupial mammals suffered heavily from the Cretaceous extinction.*

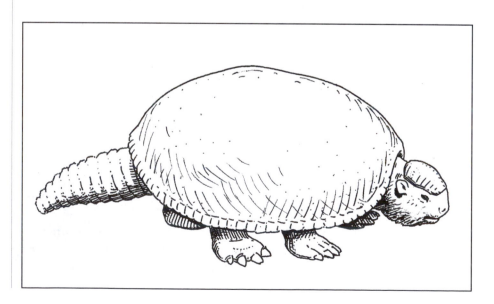

After the dinosaurs became extinct, the mammals underwent an explosive evolutionary radiation that gave rise to several unusual species. Many of these, because of their overspecialization, became evolutionary dead ends. The mammals successfully replaced the dinosaurs possibly because of their more advanced biological development, placing the mammals higher up on the evolutionary scale. This advancement might have given the mammals a decisive edge during times of environmental stress, enabling them to survive the extinction at the detriment of the dinosaurs.

Extremes in climate and topography during the Cenozoic were caused by rapid continental movements (Table 8), which built many of the world's mountain ranges. These changes provided a greater variety of living conditions than during any other equivalent span of geologic history. The rigorous environments presented the mammals with many challenging opportunities. The invasion of new habitats or the repopulation of the Earth following a mass extinction is the major source of evolutionary opportunities. As a response to those opportunities, species show a burst of evolutionary development that produces innovative adaptations.

Advanced species evolve at higher rates than those lower down on the evolutionary scale, especially marine organisms. Although extinction in the oceans at the end of the Cretaceous was severe and many species died out, very few radical species evolved as a result. Vacated ecologic niches were simply occupied by closely related species. However, a similar replacement did not occur on land because the dinosaurs represented the largest group of terrestrial animals. When they died out the world was left wide open to invasion by the mammals. In the wake of the extinction, a wide range of ecologic niches suddenly became vacant. The group that could diversify the fastest and rapidly fill those vacancies was the most successful.

THE FOSSIL RECORD

Paleontology is a branch of geology devoted to the study of ancient life based on fossil evidence. Fossils are the remains or traces of extinct organisms preserved from the geologic past. A major problem encountered when exploring for fossils of early life, however, is that the Earth's crust is constantly changing. Only a few fossil-bearing formations have survived undisturbed over time. Furthermore, not all species fossilize. Plants and animals must be buried under rigorous conditions for fossilization to occur. When given the right circumstances and sufficient time, the remains of organisms are modified, petrified, and literally turned to stone (Fig. 92). When placed into their proper order, fossils pieced together a nearly complete historical account of life on Earth, showing clear evidence for the evolution and extinction of species.

	Geologic division (Millions of Years)	Gondwana	Laurasia
TABLE 8		**CONTINENTAL DRIFT**	
Quaternary	5		Opening of the Gulf of California
Pliocene	11	Spreading begins near the Galápagos Islands	Spreading changes directions in the eastern Pacific
		Opening of the Gulf of Aden	Birth of Iceland
Miocene	26		
		Opening of the Red Sea	
Oligocene	37		
		Collision of India with Eurasia	Spreading begins in the Arctic Basin
Eocene	54		Separation of Greenland from Norway
Paleocene	65	Separation of Australia from Antarctica	
			Opening of the Labrador Sea
		Separation of New Zealand from Antarctica	
			Opening of the Bay of Biscay
		Separation of Africa from Madagascar and South America	Major rifting of North America from Eurasia
Cretaceous	135		
		Separation of Africa from India, Australia, New Zealand, and Antarctica	Separation of North America from Africa begins
Jurassic	180		
Triassic	250	Assembly of all continents into the supercontinent Pangea	

One of the most practical uses of fossils is in delineating rock formations for geologic mapping. The purpose of geologic maps is to display the distribution of rocks on the Earth's surface. They present in plan view the geologic history of an area where particular rock exposures are found. The maps also

indicate the relative ages of these formations and profile their position underground. Often, much of the information is compiled from just a few available exposures, which must be extrapolated over a large area.

The significance of fossils as a geologic tool was discovered in the late 18th century by the English civil engineer William Smith. He found that rock formations in the canals he built across Britain contained fossils significantly different from those in the beds above or below. He also noticed that sedimentary strata in widely separated areas could be identified by their distinctive fossil content. Therefore, layers from two different sites could be regarded as equivalent in age as long as they contained the same fossils. Furthermore, a bed such as sandstone might grade into another bed such as limestone that contains the same fossils, indicating the strata are the same age.

Smith drew geologic maps of the varied rock formations throughout Britain by using the characteristics of the different strata and their fossils. He made the most significant contribution to the understanding of fossils when he proposed the law of faunal succession. This law stated that rocks could be placed into their proper time sequence by studying their fossil content. It became the basis for the establishment of the geologic time scale and the beginning of modern geology.

Figure 92 *A petrified tree trunk in the Chinle Formation, Petrified Forest National Park, Apache County, Arizona.*

(Photo by A. A. Baker, courtesy USGS)

The French geologists Georges Cuvier and Alexandre Brongniart refined this approach with their discovery that certain fossils were confined to specific beds. The geologists arranged fossils in a chronologic order and noticed a systematic variation according to their positions in the geologic column. Fossils in the higher rock layers more closely resembled modern species than those farther down. The fossils did not occur randomly but in a determinable order from simple to complex. Units of geologic time could therefore be identified by their distinctive fossil content.

British geologist Charles Lyell took these ideas one step further in 1830. He proposed that rock formations and other geologic features took shape, eroded, and reformed at a constant rate throughout time according to the theory of uniformitarianism. This theory essentially states that the present is the key to the past. In other words, the forces that shaped the Earth are uniform and operated in the past much the same as they do today. The theory was originally developed in 1785 by Lyell's mentor, the Scottish geologist James Hutton, who is known today as the "father of geology." Hutton recognized the slow processes by which geologic forces operated. His discovery of unconformities, where ancient sedimentary strata were upturned, eroded, and blanketed by younger deposits (Fig. 93), suggested the history of the Earth is exceedingly

Figure 93 *Quaternary terrace deposits resting unconformably on eroded edges of Monterey shale, San Luis Obispo County, California.*

(Photo by G. W. Stose, courtesy USGS)

long and complex. Hutton believed contemporary rocks at the surface were formed by the waste of older rocks that was laid down in the sea, consolidated under great pressure, and upheaved onto dry land, as indicated by fossils embedded in the strata.

The arrangement of fossils according to their age does not present a random or haphazard picture. Instead, it shows progressive changes in life-forms and reveals the advancement of species through time. Geologists could thus recognize geologic periods based on groups of organisms that were especially plentiful and characteristic during a particular interval. Within each period, the occurrence of certain species determined several subdivisions. This succession exists on every major continent and is never out of order.

Geologic time is measured by tracing fossils through the rock strata and noting the greater change with rocks farther down the geologic column compared with those near the surface. Fossils are necessary for correlating rock units over vast distances due to a changing lithology. By using fossils, geologists could match rocks between widely separated areas or between continents. Fossil-bearing strata could thus be followed horizontally over great distances because a particular fossil bed is identifiable in another locality with respect to beds above and below it. These are called marker beds and are important for making rock correlations.

Major breaks in geologic time often occur at points of mass extinctions, when life was forced to start anew. Global catastrophes, such as mass extinctions, are normal occurrences in geologic history. Therefore, extinctions play an important role in the uniformitarian process, which postulates that the slow changes observed on Earth today had their counterparts in the past. Certain periods of mass extinctions appear to contradict this notion, however. They result from catastrophic events rather than the usual modes by which extinctions occur such as subtle changes in climate or sea level or an increase in predation.

Fossils of extinct species are used to establish a geologic time scale that can be applied to all parts of the world. The geologic time scale based on the appearance and disappearance of species was devised by 19th-century geologists. The history of the Earth has been divided into units of geologic time according to the type and abundance of fossils present in the strata. The periods take their names from the localities with the best exposures (Fig. 94). For example, the Jurassic period is named for the Jura Mountains in Switzerland, whose limestones provide a suit of fossils that reasonably depicts the period.

Both large and small extinctions define the boundaries of the geologic time scale. However, because no means was available to date rocks directly, the entire geologic record was delineated by using relative dating techniques. These indicated only which bed was older or younger according to its fossil content. Therefore, relative dating only places rocks into their proper sequence or order but does not indicate how long ago an event took place. Relative dating simply shows that one event followed or preceded another event. With the development of radiometric dating techniques, geologists had a method of precisely dating geologic events.

Rock formations are classified into erathems, which consist of the rocks formed during an era of geologic time. Erathems are divided into systems which consist of rocks formed during a period of geologic time. Systems are divided into groups, which consist of rocks of two or more formations that contain common features. Formations are classified by distinctive features in the rock and are given the name of the locality where they were originally described. Formations are divided into members, which might be subdivided into individual beds such as sandstone, shale, or limestone.

A type section is a sequence of strata that was originally described as constituting a stratigraphic unit and serves as a standard of comparison for identifying similar, widely separated units. Preferably a type section is selected in an area where both the top and bottom of the formation are exposed. Type sections are named for the area where they are best exposed. For example, the Jurassic Morrison Formation, well-known for its dinosaur bones (Fig. 95), is named for the town of Morrison outside Denver, Colorado. Type sections are also distinguished by their distinctive fossil content, which is used to correlate

Figure 94 *Type localities for the various geologic periods.*

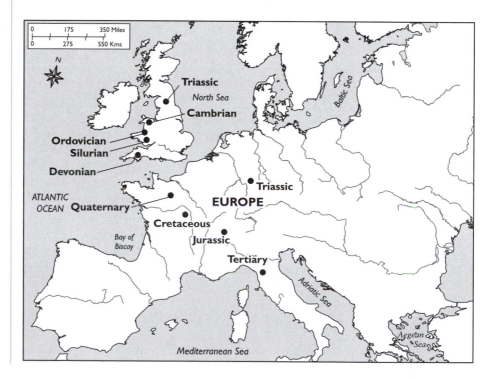

Figure 95 *The Jurassic Morrison Formation in the Uinta Mountains, Utah.*

(Author's collection)

stratigraphic units. These are placed in sequence by age into a geologic column and are used to establish a geologic time scale.

After seeing how mass extinctions affect the development of life, the next chapter will examine the processes that affect the evolution of species, including natural selection, genetic mutations, and the survival of the fittest.

7

EVOLUTION OF SPECIES
THE ADVANCEMENT OF LIFE

This chapter examines the factors that affect the evolutionary development of species. The two fundamental types of classical evolutionary evidence consist of information on organisms living today along with fragmentary evidence on earlier life-forms as read in the fossil record. The history of life as told by fossils is not complete, however, because of the constant remaking of the Earth's surface, which erases whole chapters of geologic history.

Yet the study of fossils along with the radiometric dating of rocks that contain them has constructed a reasonably good chronology of historical geology. Today, geologic history is dated satisfactorily for most purposes. Future discoveries will adjust the geologic time scale to better perfection. Even in its present form, the fossil record provides clear evidence for the evolution of life on Earth.

BIOLOGIC DIVERSITY

One of the most striking and consistent patterns of life on this planet is the greater the profusion of species when moving farther from the poles and closer to the equator. This occurs because near the equator, more solar energy is available for photosynthesis by simple organisms, the first link in the global food chain. Other factors that enter into this energy-species richness relationship include the climate, the available living space, and the geologic history of the region. For instance, coral reefs and tropical rain forests support the largest species diversity because they occupy areas with the warmest climates.

The world's oceans also have a higher level of species diversity than the continents. Due to a lower ecologic carrying capacity, which is the number of species an environment can support, the land has limited the total number of genera of animals since they first crawled out of the sea some 350 million years ago. The marine environment, by comparison, supports twice the living animal phyla of the terrestrial environment. Marine biologic diversity is influenced by ocean currents, water temperature, the nature of seasonal fluctuations, the distribution of nutrients, the patterns of productivity, and many other factors of fundamental importance to living organisms.

The vast majority of marine species live on continental shelves or shallow-water portions of islands and subsurface rises generally less than 600 feet deep. The richest shallow-water faunas are at low latitudes in the tropics, which contain large numbers of highly specialized species (Fig. 96). Shallow-water environments also tend to fluctuate more than those farther offshore, which significantly affects evolutionary development.

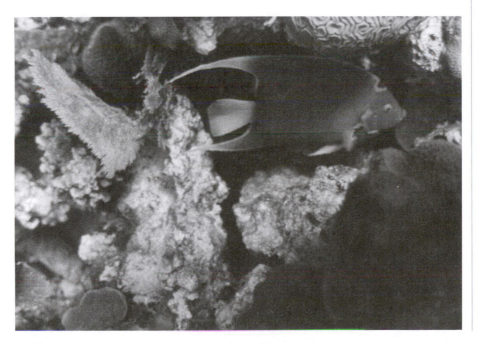

Figure 96 *An angelfish swims among rock and coral near High Bay, Andros Island, Bahama Islands.*

(Photo by P. Whitmore, courtesy U.S. Navy)

Species diversity progressively falls off in the higher latitudes. Less than 10 percent as many marine organisms live in the polar regions than in the tropics. Moreover, the Arctic Ocean, which is surrounded by continents, supports twice the biologic diversity as the Antarctic Ocean, which surrounds a large continent. Species living at high latitudes occupy broad geographic ranges, whereas those crowding the tropics are much more restricted.

Biologic diversity depends mostly on the stability of the food supply. This is largely affected by the shape of the continents, the extent of inland seas, and the presence of coastal mountains. Erosion of mountains pumps nutrients into the sea, fueling booms of marine plankton and increasing the food supply for animals higher up the food chain. Organisms with abundant food are more likely to thrive and diversify into different species. Mountains that arise from the seafloor to form islands increase the likelihood of isolation of individual animals and, in turn, increase the chances of new species forming.

Continental platforms are particularly important because extensive shallow seas provide a large habitat area for shallow-water faunas and tend to dampen seasonal climatic variations. This makes the local environment more hospitable. As the seasons become more pronounced in the higher latitudes, food production fluctuates considerably more than in the lower latitudes. Species diversity is also influenced by seasonal changes such as variations in surface and upwelling ocean currents (Fig. 97). These affect the nutrient supply and thereby cause large fluctuations in productivity.

Figure 97 Upwelling and sinking ocean currents off seacoasts affect marine productivity.

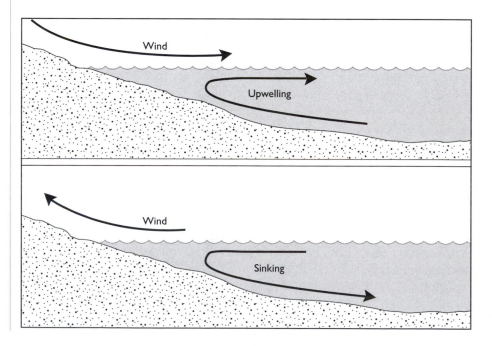

Figure 98 *A fish swims near the sea bottom at Point Loma off of San Diego, California.*

(Photo by R. Outwater, courtesy U.S. Navy)

The highest diversity is off the shores of small islands or small continents in large oceans, where fluctuations in nutrient supplies are least affected by the seasonal influences of landmasses. The lowest diversity is off large continents, particularly when they face small oceans, where shallow-water seasonal variations are the greatest. Diversity also increases with distance from large continents.

The coral reef environment supports more plant and animal species than any other marine ecosystem. Tropical plants and animals thrive on the reefs, taking advantage of the corals' ability to build massive, wave-resistant structures. The key to this prodigious growth is the unique biology of corals, which plays a critical role in the structure, ecology, and nutrient cycles of the reef community. Coral reef environments have the highest rates of photosynthesis, nitrogen fixation, and limestone deposition of any habitat. The most remarkable feature of coral colonies is their ability to build massive calcareous skeletons weighing several hundred tons.

Living on this framework are small, fragile corals and large communities of green and red calcareous algae. Hundreds of species of encrusting organisms such as barnacles thrive on the coral framework. Large numbers of invertebrates and fish hide in the nooks and crannies of the reef (Fig. 98), often waiting until nighttime before emerging to feed. Other organisms attach to practically all available space on the underside of the coral platform or on dead coral skeletons. Filter feeders such as sponges and sea fans occupy the deeper regions.

Species living in different oceans or on opposite sides of the same ocean evolve separately from their overseas relatives. Even along a continuous coast-line, major changes in species generally correspond to the climate because latitudinal and climatic changes create barriers to shallow-water organisms. The deep-sea floor provides another formidable barrier to the dispersal of shallow-water organisms.

These barriers partition marine faunas into more than 30 geographic provinces with only a few common species living in each province. The shallow-water marine faunas represent more than 10 times as many species than found in a world with only a single province. For example, around 200 million years ago, when a single large continent named Pangaea was surrounded by a global ocean, marine species diversity was at an all-time low.

The widest ranging and the most diverse of all marine provinces is the Indo-Pacific due to its long chains of volcanic island arcs (Fig. 99). When island chains align east to west within the same climate zone, they attract highly diverse, wide-ranging faunas. These spill over onto adjacent tropical continental shelves and islands. This vast tropical biota is isolated from the

Figure 99 Long chains of islands in the Indo-Pacific attract diverse, wide-ranging faunas.

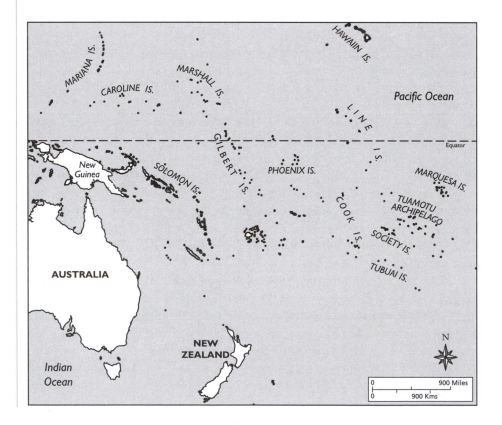

TABLE 9   CLASSIFICATION OF ORGANISMS

Kingdom (plants, animals, archea)
      Phylum (33 phyla)
         Class
            Order
               Family
                  Genus (average 60 species per genus)
                     Species (1.75 million known species)
                        Breed (groups of closely related organisms)

---

the evolutionary scale becomes more inclusive, encompassing a larger number of organisms.

The science of cladistics reformed classification by using only the branching order of lineages on evolutionary trees. Therefore, in the cladistic system, lungfish are more closely related to land vertebrates than to bony fish even though they are more similar to the latter in outward appearance. This occurs because the common ancestor of lungfish and terrestrial vertebrates appears more recently in time than that linking the lungfish with modern bony fish. In other words, the cladistic system of classification works only with the closeness of common ancestry in time without regard to similarities in morphology or appearance.

In the present classification scheme, each organism is assigned an italicized, two-part genus and species name. The first word, which is capitalized and italicized, is the generic name and is shared by other closely related species. The second word, written in lowercase and italicized, is the species name and is unique to a particular genus. For example, humans' species name is *Homo sapiens*. The genus *Homo* also applies to our human predecessors such as *Homo habilis* and *Homo erectus*.

Sometimes the discoverer's name and the date of discovery follow the species name. The scientific name of a species is written either in Latin or Greek to provide enough names for some 2 million known species and the hundred or so new specimens discovered daily. All species must have a unique name that cannot be used again for another species. Nor can more than one name be applied to the same species.

A kingdom is the largest taxonomic classification and contains all types of animals, plants, or microorganisms. The kingdom humans belong to is Animalae, which comprises all animals. The phylum Chordata comprises all vertebrates. The class Mammalia comprises all mammals. The order Primate also

western shores of the Americas by a long stretch of c
effective barrier to the migration of shallow-water ma

Biologic diversity depends mostly on the food s
organisms called phytoplankton are responsible for more
all marine photosynthesis. They play a critical role in th
which spans 70 percent of the Earth's surface. Phytoplankto
producers in the ocean and occupy key positions in marine l
organisms residing at the very bottom of the food web are eat
which are preyed upon by progressively larger predators. Plan
duce 80 percent of the breathable oxygen as well as regulate ca
which largely controls the global climate.

The ocean's color varies markedly, depending on suspended
as phytoplankton, sediments, and pollutants. In the open ocean,
biomass is low, the water has a characteristic deep-blue color. In th
ate coastal regions, where the biomass is high, the water has a c
greenish tinge. The temperate waters of the North Atlantic are tinte
simply because they contain a rich supply of phytoplankton.

Scattered around the world are numerous upwelling zones oi
nutrient-rich water that support prolific booms of phytoplankton and
marine life. Upwelling ocean currents off the coasts of continents and
the equator bring to the surface important sources of bottom nutric
such as nitrates, phosphates, and oxygen. These regions cover only abou
percent of the ocean, yet they account for some 40 percent of all marir
productivity.

SPECIES CLASSIFICATION

Less than 2 million of the estimated 10 to 30 million extant, or living, species
have been formally classified. Paleontologists use the same system to classify
extinct species as they do for classifying extant organisms. The 18th-century
Swedish botanist Carl von Linné, better known under his latinized name Car-
olus Linnaeus, proposed the first classification scheme. He based his classification on the number of characteristics organisms had in common. He named
organisms using Latin words because this was the universal language of science
in his day.

Linnaeus realized that some organisms shared similarities simply because
they were more closely related. Later, as evolution became recognized as the
process by which organisms develop into new species, classification schemes
were devised to take into account these evolutionary patterns, demonstrating
how groups of organisms were related both in space and time. The classification of taxa (taxonomic groups) establishes a hierarchy (Table 9). Each step up

comprises monkeys and apes. The family Hominidea comprises all humans. Genus and species are the lowest rungs of the classification hierarchy. The genus *Homo* also includes humans' primitive ancestors, beginning with *Homo habilis,* which lived about 2 million years ago and was the first human to make tools. However, the species name *sapiens,* used in *Homo sapiens,* applies only to extant humans.

Of the 33 existing phyla, only 10 are necessary to classify the vast majority of animal life on Earth from both the past and the present (Table 10). The first phylum, Protozoa, begins with the simplest organisms (Fig. 100). Each succeeding phylum becomes more complex. The phylum Chordata, where humans belong, has the most complexity. Thus, the ordering of phyla in this manner recognizes the evolutionary advancement of species.

All phyla of advanced animals, with the possible exception of Bryozoa, had their beginning in the Cambrian explosion, when a variety of fundamentally new body types appeared in a geologic moment. Little has happened since in terms of novel basic body forms. This is because once evolution fills the world with sufficient variety, further innovation is no longer needed. Marine animals have only so many unique ways of feeding,

TABLE 10 CLASSIFICATION OF SPECIES

Group	Characteristics	Geologic Age
Vertebrates	Spinal column and internal skeleton. About 70,000 living species. Fish, amphibians, reptiles, birds, mammals.	Ordovician to recent
Echinoderms	Bottom dwellers with radial symmetry. About 5,000 living species. Starfish, sea cucumbers, sand dollars, crinoids.	Cambrian to recent
Arthropods	Largest phylum of living species with over 1 million known. Insects, spiders, shrimp, lobsters, crabs, trilobites.	Cambrian to recent
Annelids	Segmented body with well-developed internal organs. About 7,000 living species. Worms and leeches.	Cambrian to recent
Mollusks	Straight, curled, or two symmetrical shells. About 70,000 living species. Snails, clams, squids, ammonites.	Cambrian to recent
Brachiopods	Two asymmetrical shells. About 120 living species.	Cambrian to recent
Bryozoans	Moss animals. About 3,000 living species.	Ordovician to recent
Coelenterates	Tissues composed of three layers of cells. About 10,000 living species. Jellyfish, hydra, coral.	Cambrian to recent
Porifera	The sponges. About 3,000 living species.	Proterozoic to recent
Protozoans	Single-celled animals. Foraminifera and radiolarians.	Precambrian to recent

Figure 100 *Fossil foraminifera of the North Pacific Ocean.*

(Photo by B. P. Smith, courtesy USGS)

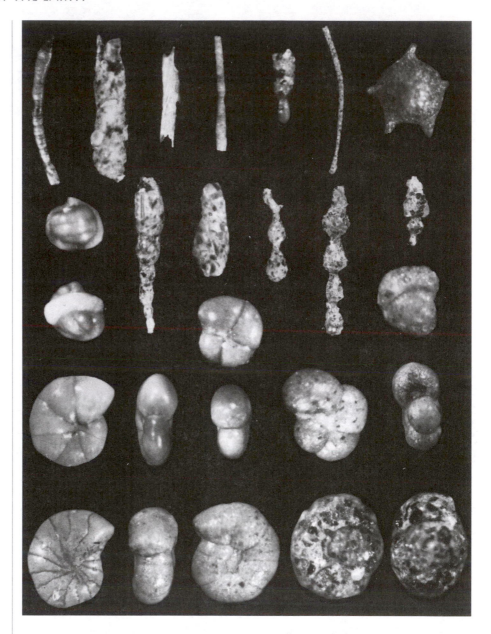

breeding, attacking, and defending themselves. Even after the Permian extinction, when species diversity was at an all-time low, few novel body plans came into existence. At that time, most ecological niches had already been filled. This was unlike the early Cambrian, when the world was ripe for invasion of new life-forms.

NATURAL SELECTION

During the global exploration of the oceanographic research vessel HMS *Beagle* from 1831 to 1836 (Fig. 101), a young British naturalist by the name of Charles Darwin was employed without pay as the ship's geologist. He described in great detail the rocks and fossils he encountered on his journey. Darwin got his ideas about evolution by poring over voluminous notes he took on his trip abroad and from direct observations made in Britain. Interestingly, Darwin was trained and worked as a geologist. However, today he tends to be viewed as a biologist. He made many significant contributions to the field of geology, which during his day was entering its golden age.

When Darwin visited the Galapagos Islands in the eastern Pacific, he noticed major changes in plants and animals living on the islands compared with their relatives on the adjacent South American continent. Animals such as finches and iguanas assumed distinct but related forms compared with those on adjacent islands. Cool ocean currents and volcanic rock made the Galapagos a much different environment than Ecuador, the nearest land 600 miles to the east. The similarities among animals of the two regions could only mean that Ecuadorian species colonized the islands and then diverged by a natural process of evolution.

Darwin proposed the theory of natural selection as the main force driving evolution, which is the mechanism producing the diversity of life. All organisms evolve through the preservation of genetically based traits that best contribute to the survival and reproduction in a particular environment. The processes of evolution within the framework of natural selection is essentially

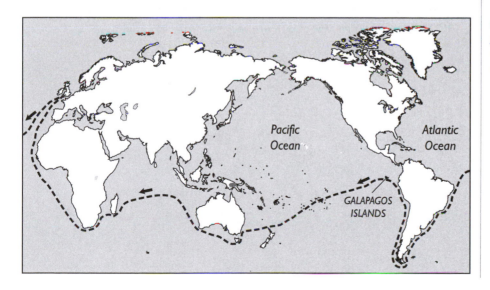

Figure 101 Darwin's journey around the world during his epic exploration.

an ecological mechanism based on relationships between organisms and their environments. Certain inherited traits allow species to be particularly well suited to survive and reproduce in their prevailing environment. The larger the number of different environments, the wider the variety of species. During environmental stress, species that acquire favorable traits through mutations adapt more easily. They are therefore more likely to survive and pass on these survival characteristics to their offspring.

Evolutionary trends varied throughout geologic time in response to major environmental changes. Natural selection acted to adapt organisms to the new conditions forced upon them by several environmental factors such as chemical changes in the ocean, climate changes, or mass extinctions. However, natural selection is not deterministic. Variations are purely accidental and are selected according to the demands of the environment. Most of the time, species resist change even though the consequences make them better suited to environmental needs.

Just as the lobe-finned fish that gave rise to the amphibians left the ocean to seek opportunities on land some 350 million years ago, so did reptiles and mammals seek similar opportunities. However, this time they went in the opposite direction—back to the sea. Several reptilian species went to sea, including lizards and turtles that were quite primitive. However, only the smallest turtles survived the extinction at the end of the Cretaceous.

The extinct mesosuchians (Fig. 102), Mesozoic relatives of the crocodiles, practiced a form of "reverse evolution," by adapting to a predator's life in the ocean. Actually, reverse evolution is a misnomer. The fossil record implies that evolution proceeds in only one direction—from simple to complex organisms. With the possible exception of certain parasites, which devolved from higher life-forms only by discarding their now useless digestive

Figure 102
Mesosuchians were Mesozoic relatives of crocodiles.

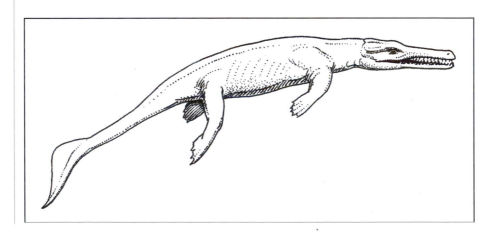

systems, no organism has been shown to degenerate to a lower form. Parasites are so prolific they comprise about two-thirds of all species.

When reptiles and mammals returned to the sea, they assumed the appearance of fish. This is called convergent evolution, in which different organisms share similar physical attributes because they occupy the same environment and have similar lifestyles. Why reptiles and mammals reentered the ocean to compete with fish is unknown. Generally, though, strong selective pressures, such as an abundance of food and habitat or intense competition, will drive species into different environments. Before these changes can occur, the organisms must be equipped with the right body parts to survive in their new habitat. The lobe-finned fish developed the proper appendages that enabled it to "swim" across the land. Although it could barely fly, *Archaeopteryx* was at least on the right track toward modern birds.

To understand evolution on a small scale requires the observation of species in transition. In central England during the 19th century, prior to the advent of coal-burning factories, many species of white moths were camouflaged against light-colored tree trunks and sufficiently protected from bird attacks. Occasional mutations also produced darker moths, which were more visible to birds and therefore at a disadvantage. During the industrial revolution, as coal soot blackened the tree trunks, the darker moths were less conspicuous while the lighter moths became easy prey. In a mere half century, a complete role reversal resulted in a new race of dark moths with a distinct advantage over their lighter-colored relatives.

Similarly, unwanted pests and weeds evolve resistance to insecticides and herbicides almost as fast as new chemicals are developed to eradicate them. Vegetation growing near industrial plants has evolved forms resistant to toxic chemicals and heavy metals polluting the soil. As people alter the chemistry of the environment with deadly pollutants, they are inadvertently causing the evolution of new and potentially dangerous species.

The giraffe is often cited as a classic example of evolutionary adaptation (Fig. 103). Its long neck evolved in response to a diet of tree leaves beyond the reach of shorter browsers. A long neck is also an effective weapon for fighting off male contenders for females. An elephant's trunk is another adaptation for browsing. In this case, however, the snout was elongated for reaching tall branches and for digging water holes during dry spells. Two living varieties of elephants presently exist in Africa and India and are closely related to the extinct woolly mammoth. Mastodons were also related and diverged from the elephant family about 30,000 years ago, only to become extinct 20,000 years later.

Elephants might have originally evolved as seagoing mammals that used their trunks as snorkels, which would account for their unusual anatomy. Elephant fetuses show kidney features not seen in any mammal that gives birth

Figure 103 *The evolution of the giraffe is clear evidence of adaptation from a grazing to a browsing lifestyle.*

to live young but, instead, that resemble those of the embryos of egg-laying mammals, such as the platypus—an aquatic monotreme. The trunk also appears early in fetal growth, suggesting a more ancient origin. Fossil bones as well as genetic studies suggest that elephants are closely related to manatees, aquatic mammals often referred to as sea cows. However, studies of the oldest elephant fossils indicate elephants originally lacked trunks.

Horses show a clear evolutionary advancement in the fossil record. The earliest horses, called eohippus (Fig. 104), originated in western North America during the Eocene some 50 million years ago. At that time, they were only about the size of a small dog. As horses progressed in size, their faces and teeth elongated as the animal switched from browsing to grazing, at which time their toes fused into hoofs. Many types of hoofed animals called ungulates evolved in response to the spread of grasslands throughout the world. Camels originating in North America about 25 million years ago migrated to other parts of the world by connecting land bridges (Fig. 105).

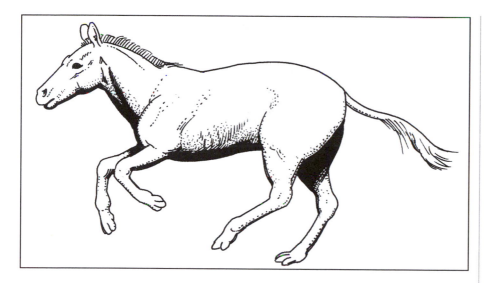

Figure 104 *Eohippus was among the earliest horses.*

GENETIC MUTATION

The means by which organisms adapt to their changing environment is through genetic mutation. Cell division called mitosis, whereby the nucleus divides into equal sets of chromosomes, increases the likelihood of genetic variation and accelerates the rate of evolution as organisms encounter new

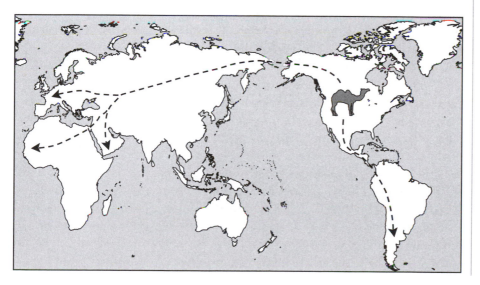

Figure 105 *The migration of camels from North America to other parts of the world.*

environments. The extraordinary variety of plant and animal life that has arisen on this planet over the last 600 million years is in response to genetic mutation and its enormous potential for species diversity. Moreover, the most successful organisms were those that managed to split into many viable species.

Altered genes in the chromosomes enact certain changes that enable a species and its offspring to live more successfully in their respective environments. Mutation rates vary according to species but generally have a frequency of about one mutation per million years. They occur as a direct result of the chemistry of the environment and the bombardment of genes by cosmic and background radiation. Thus, the environment directly affects the rate of mutation. The mutations can be either beneficial or harmful and therefore dictate whether an organism survives or goes extinct. Not until ecosystems are perturbed and species start going extinct at high rates, however, will evolution occur, producing new adaptations and new species to rebuild the disrupted habitats.

Rapid evolutionary advancements might result from rare, large mutations. Thus, evolution appears to make sudden leaps, with major changes occurring simultaneously in many body parts. In other words, natural selection does not favor partial development. Natural selection therefore cannot work on structures that are not fully functional during intermediary periods of development of new appendages. An example is the development of insect wings, which were probably first used for cooling purposes. Later, as the benefits of flight became apparent, the wing structures grew more aerodynamic, giving flying insects an enormous advantage over their earthbound competitors.

Apparently, the only organisms that can survive radical changes in the environment, such as abrupt temperature shifts or food shortages, are those that have mutated before encountering the stress. Those species that altered into a more adaptive form before the upheaval were better able to withstand sudden environmental stress. For example, diatoms with siliceous cell walls (Fig. 106) survived the extinction that ended the Cretaceous far better than other single-celled plankton possibly because they evolved a dormancy mechanism for weathering seasonal droughts during normal times. The diatoms were simply fortunate to possess a favorable feature that evolved for other purposes.

Nevertheless, experiments on bacteria might force a revision of this view. Bacteria can adopt genetic traits in response to a particular environment. They then pass on these acquired traits to their offspring, providing them with a better chance for survival. Mutations in the bacteria arise spontaneously and randomly, giving the organism the ability to mutate in a more purposeful manner so as to adapt to a particular environment. Unfortunately, potentially

Figure 106 *Marine diatoms from the Choptank Formation, Calvert County, Maryland.*

(Photo by G. W. Andrews, courtesy of USGS)

dangerous bacteria have also evolved forms that are completely resistant to penicillin and other antibacterial drugs.

Mass extinctions select for survival those individuals that have altered in advance into a better adaptive form. They are not the direct cause of the mutations, however. Therefore, certain species can survive one mass extinction after another. This is particularly true for marine species such as sharks, which have weathered every mass extinction over the last 400 million years. Regrettably, many shark species are presently facing extinction from their fiercest predator: humans.

SURVIVAL OF THE FITTEST

On his voyage around the world, Darwin observed the relationships between animals on islands and on adjacent continents as well as between extant animals and fossils of their extinct relatives. These studies led him to conclude that species constantly evolved throughout time. Therefore, according to Darwin, evolution worked at a consistent rate as species adapted to a continuously changing environment.

Darwin was not the first to make this observation, however. Where his theory differed was in postulating that new parts evolved in minute stages rather than in discrete jumps. Darwin attributed such apparent rapid evolution to an artifact in the fossil record by which erosion or nondeposition of fossil-bearing strata caused gaps in geologic time.

Darwin coined the phrase "survival of the fittest," meaning species best suited to their environment have a better chance of producing offspring that possess the survival characteristics of their parents. In terms of strict Darwinism, a constant struggle for reproductive success occurs among organisms within a population. Successful parents pass on their "good" genes to their offspring, making them better able to survive in their respective environments. In this manner, natural selection weeds out the weak in favor of more adaptable species.

Contemporary geologists embraced Darwin's theory for it offered a clear understanding of the changes in evolutionary development observed in fossils of different ages. Only fossils of extinct species can testify to the historic patterns of evolution traced out through geologic time. Geologists could thus place geologic events into their proper sequence by studying the evolutionary changes taking place among fossils.

Evolution was not always gradual and of constant tempo as Darwin believed. The fossil record implies that life evolved erratically, with long periods of little or no change punctuated by short intervals of rapid change and then followed by long periods of stasis (stay-as-is). The pattern of change and stasis is called punctuated equilibrium. New species evolved within a few thousand years, practically instantaneously in geologic time, and then remained essentially unchanged for up to several million years.

Disparate species survive together through millions of years of environmental change and then disappear en masse during particularly abrupt upheavals in a pattern known as coordinated stasis. This runs counter to traditional evolutionary theory whereby species evolve on their own. Communities of organisms might become linked in such a way that helps them survive some types of environmental disturbances. However, if sea level, climate, or some other environmental factor changes too radically, the community loses its cohesion and species go extinct.

Not until ecosystems are perturbed and species start going extinct at high rates will evolution take control, producing novel adaptations that give rise to new species. Little happens in evolution without extinction first disrupting ecosystems and driving many preexisting stable species extinct. The extinction nearly always results from the physical environment changing beyond the point where species can relocate by finding familiar habitat elsewhere. The more devastating and globally encompassing an extinction event, the greater the evolutionary change.

Species formed relatively quickly as a result of rapid bursts of evolutionary change. Furthermore, rapid evolutionary changes in large segments of organisms might appear in the fossil record as though caused by mass extinction when, in fact, no actual extinction had occurred. Evolution by natural selection might also be opportunistic, with variations arising by chance and selected according to the demands of the environment.

Extinction survival, therefore, might simply be the luck of the draw as winners replace losers. The evolution of a single species also affects the evolution of others with which it interacts. When the environment changes abruptly to one that is more demanding, species incapable of rapidly adapting to these new conditions cannot live at their optimum and therefore do not pass on their "bad" genes to future generations.

MISSING LINKS

The history of the Earth is incomplete because of gaps in the fossil record. Entire chapters are missing due to periods of erosion or nondeposition of sedimentary strata that traps and preserves species as fossils (Fig. 107). The vast majority of fossils are those of marine animals because seawater is a good preservative and sedimentation takes place more readily in water. Furthermore, marine life has been around in one form or another about eight times longer than terrestrial life, providing a greater number of species available for fossilization.

Gaps in the fossil record might also result from insufficient intermediary species, or so-called "missing links." These apparently existed only in small populations. Small populations are less likely to leave a fossil record because the process of fossilization favors large populations. The intermediates probably did not live in the same locality as their ancestors and thus were unlikely to be preserved along with them. New species that originate in small populations seem to evolve rapidly as they radiate into new environments. Then, as populations increase, slower evolutionary changes occur as the species' chances of entering the fossil record improve.

The origination and extinction of species throughout geologic history are important for building an accurate account of the evolution of species

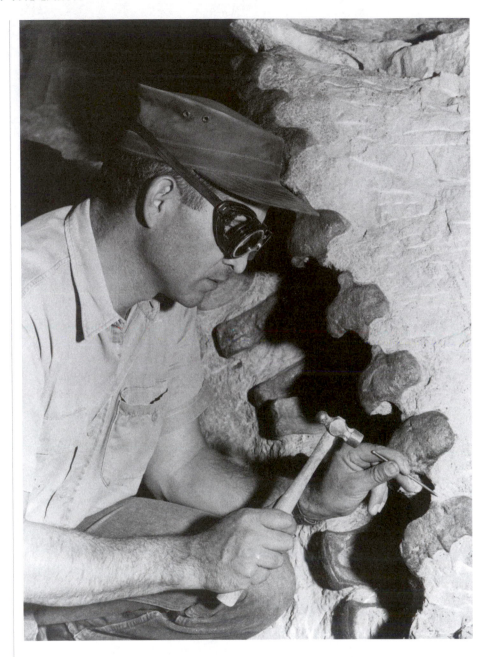

Figure 107 *A paleontologist carefully chips away rock from a column of dinosaur vertebrae at Dinosaur National Monument, Uinta County, Utah.*

(Photo by W. R. Hansen, courtesy USGS)

through time. However, ambiguities in the fossil record suggest differences in fossil samples where no actual differences exist. In any ecologic community, a few species occur in abundance, some occur frequently, but most are rare and occur only infrequently. The chances of individuals becoming fossilized after death and thus entering the fossil record are extremely small.

Another problem is the method of sampling. For instance, no single fossil sample will contain all the rare organisms in an assemblage of species. By comparing one sample with another higher up in the stratigraphic column, which represents a later time in geologic history, an overlapping but different set of rare species occurs. Species found in the lower sample but missing in the upper sample might therefore be erroneously inferred to have gone extinct. Conversely, species appearing in the upper sample but not found in the lower sample might be wrongly thought to have originated there. Thus, the reading of the fossil record can often be confusing and misleading.

THE GAIA EFFECT

Generally, environmental changes are thought to drive evolution. However, in 1979, the British chemist James Lovelock turned this argument around by proposing the Gaia hypothesis, named for the Greek goddess of the Earth. It postulates that life is largely in control of its environment and that the biosphere maintains the optimum living conditions for all organisms by regulating the climate. For example, a certain species of plankton releases into the atmosphere a sulfur compound that aids in the formation of clouds. During a warmer climate, plankton growth is invigorated, releasing more cloud-forming sulfur compounds. The increased cloud cover cools the planet and reduces plankton growth, thus offering an effective feedback-control mechanism that stabilizes the global temperature.

The Gaia hypothesis suggests that from the very beginning, life followed a well-ordered pattern of growth, advancing from simple to complex organisms independent of chance and natural selection. Life appears to have kept pace with all major changes in the Earth over time and made significant alterations of its own. One major change was the conversion from an abundance of carbon dioxide in the early atmosphere and ocean to a high concentration of oxygen today due to photosynthesis (Table 11).

As life progressed, slow but steady changes took place that greatly affected the final outcome of the planet. Like the Earth, the other planets and their satellites are comprised of a core, a mantle, a crust, and even an atmosphere and possibly a liquid hydrosphere. Only the Earth, however, possesses a biosphere. Moreover, a biosphere consists of more than just living entities. Life also must be integrated with the lithosphere, hydrosphere, and atmosphere to constitute a fully developed biosphere (Fig. 108).

Since the beginning, life has responded to a variety of chemical, climatological, and geographic changes in the Earth, forcing species to either adapt or perish. Many dead-end streets along the branches of the evolutionary tree exist in the fossil record, which itself represents only a fraction of all species

TABLE 11   EVOLUTION OF THE BIOSPHERE

Events	Billions of Years Ago	Percent Oxygen	Biological Effects	Results
Full oxygen conditions	0.4	100	Fish, land plants, and animals	Approach present biological environs
Appearance of shelly animals	0.6	10	Cambrian fauna	Burrowing habitats
Metazoans appear	0.7	7	Ediacaran fauna	First metazoan fossils and tracks
Eukaryotic cells appear	1.4	> 1	Larger cells with a nucleus	Redbeds, multicellular organisms
Blue-green algae	2.0	1	Algal filaments	Oxygen metabolism
Algal precursors	2.8	< 1	Stromatolite mounds	Initial photosynthesis
Origin of life	4.0	0	Light carbon	Evolution of the biosphere

Figure 108 The interaction between the lithosphere, hydrosphere, and atmosphere.

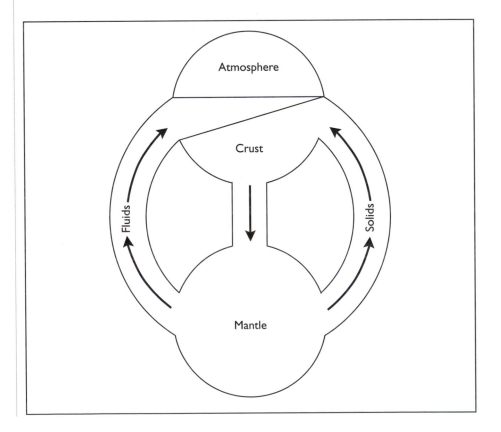

that have ever lived. Nearly every conceivable form and function have been tried, some with a higher degree of success than others. By a trial-and-error method of speciation, natural selection has chosen some species to prosper, while other, less adaptive species simply went extinct.

One of the greatest forces influencing evolutionary change is plate tectonics and the drifting of the continents. Continental motions had a wide-ranging affect on the distribution, isolation, and evolution of species. The changes in continental configuration greatly affected global temperatures, ocean currents, and productivity. The positioning of the continents with respect to each other and to the equator helped determine climatic conditions. When most of the land huddled near the equatorial regions, the climate was warm. However, when lands wandered into the polar regions, the climate grew cold and initiated glaciation.

Periods of glaciation might also have been a major component in the evolution of life on Earth. Around 680 million years ago, glacial ice 300 feet deep covered the oceans, temperatures remained well below freezing year-round, and the land was barren, dry, frigid, and lifeless. This was perhaps the planet's coldest and longest ice age. However, it might also have been a vital period in the evolution of plants and animals. The glacial period ended with the sudden appearance of complex and new life-forms. This bloom of new species might have been a key event in the long sweep of evolution that helped to create a warm, temperate planet.

After an examination of the factors that affect species evolution, the next chapter will focus on the critical cycles that influence life on Earth, ranging from the solar cycle to the geochemical cycle.

8

THE LIFE CYCLES
THE PERIODICITY OF NATURE

This chapter examines the various cycles that have influenced the growth of life on this planet since it first began. Practically every aspect of life is governed by cycles. Perhaps the periodicity of the Earth-Moon system was responsible for the origination of living entities in the first place. The constant waxing and waning of the tides accounts for the prodigious growth in the intertidal zones. The solar cycle appears to influence the climate, which dramatically affects life. The Earth's orbital cycles also seem to control the long-term climate, spawning episodes of glaciation.

The Earth has its own inner cycles that affect geologic and biological processes. The cycles might be influenced by outside forces such as the gravitational pull of the Sun, Moon, and planets. The mantle's internal rock cycle is responsible for shifting the continents around the surface of the globe on timescales of hundreds of millions of years. Increased mantle convection causes a higher incidence of volcanic activity, which in turn has a profound effect on climate and life.

THE SOLAR CYCLE

In 1894, the English astronomer Walter Maunder discovered an apparent reduction in sunspot activity between the years 1645 and 1715, named the Maunder minimum in his honor. It correlated with the coldest part of the Little Ice Age, a span of unusually cold weather in Europe and North America from the late 15th century to the middle 19th century. Global temperatures dropped 1 degree Celsius, and glaciers that had been steadily retreating since the end of the last ice age were once again advancing.

The unusual low level of solar activity is also supported by a gap in Chinese naked-eye sunspot records for the same period. Tree rings formed during this interval indicate a period of unusual behavior of the Sun, closely matching the Maunder minimum. Another Maunder-type minimum of sunspot activity occurred during the Ming dynasty in China between 1400 and 1600, which might have initiated the Little Ice Age.

Astronomers have, for centuries, referred to a solar constant, meaning the total amount of solar energy impinging on the Earth has remained steady through time. The solar constant depends on the Sun's luminosity, or brightness, and the Earth's orbit. The luminosity depends on the Sun's size and surface temperature. A reduction of the solar constant by only a few percent over an extended period could initiate glaciation.

In 1843, the German astronomer Heinrich Schwabe pointed out a pronounced cycle of sunspot activity recurring approximately every 11 years. The sunspots individually persist for only a limited time. During each cycle, they initially appear at high solar latitudes and migrate progressively closer to the solar equator, signifying the equator rotates faster than the poles. Moreover, the polarity of the Sun's magnetic field reverses every other sunspot cycle, or about 22 years. The sunspots are linked with these magnetic fields, which are several thousand times stronger than the geomagnetic field on the Earth's surface.

During a sunspot maximum, when large numbers of sunspots mar the Sun's surface, solar activity increases. During low sunspot activity, the Sun cools by as much as 1 percent. Small, short-term fluctuations in solar output are sufficient to produce variations in the climate. These minor changes occur in regular intervals of about 22 years, known as the solar cycle. Other solar activity, including solar flares (Fig. 109), solar cosmic rays, ultraviolet rays, and X rays, vary directly with the solar cycle. Coinciding with the solar cycle is an 11-year sunspot cycle. Cycles of 90 and 180 years are also prevalent. Longer-period solar cycles might correlate with the Pleistocene Ice Ages, which have returned repeatedly roughly every 100,000 years for the last 2 million years or so.

The solar cycle might also be regulated by the gravitational forces of the planets. Throughout this century, the alignment of the inner planets on one

side of the Sun appeared to control the number of sunspots. Moreover, Jupiter's orbital period nearly equals the 11-year sunspot cycle. When the planets line up together, their combined gravitational pull raises tides on the surface of the Sun. The tidal amplitudes are extremely small, however, due to the minimal gravitational attraction of the planets. Nevertheless, the years of sunspot maxima and minima since 1800 closely coincide with the Sun's tidal maxima and minima.

The alignment of the planets (Fig. 110) appears to influence the Earth's weather directly. The ancient Chinese astronomers first recognized this alignment known as a planetary synod, from the Greek *synodos* meaning "conjunction." Chinese researchers found that planetary alignments appeared to have affected the weather for the past 3,000 years. Approximately every 180 years, the Earth aligns on one side of the Sun while the other planets reside on the other side, displacing the gravitational center of the solar system. This configuration of the planets stretches the Earth's orbit nearly 1 million miles, causing the climate to cool minutely for several years. The last planetary synod began in October 1982. It could possibly initiate a cold spell lasting nearly half a century, provided other influences such as greenhouse warming do not interfere.

Volcanic eruptions are also known to follow the 11-year solar cycle, which is a slight waxing and waning in the Sun's energy output. The study of hundreds of eruptions over the past four centuries implies that the solar cycle might have had an influence on when volcanoes came to life. The eruptions appeared most numerous during the weakest portion of the solar cycle, when the number of sunspots is low. At the peak of the solar cycle, emissions from the Sun cause small but abrupt changes in the Earth's atmosphere, jarring the planet slightly. This motion might trigger tiny earthquakes that relieve stress

Figure 110 *Planets in the solar system.*

(Courtesy NASA)

163

under volcanoes, thereby preventing a large eruption until the solar cycle is again at a minimum and the pent-up volcanoes explode.

Evidence for a solar cycle operating as far back as the Precambrian is thought to exist in 680-million-year-old glacial banded deposits called varves found in lake bed sediments north of Adelaide in South Australia. The varves consist of alternating layers of silt laid down annually during the late Precambrian ice age. Each summer when the glacial ice melted, sediment-laden meltwater discharged into a lake below the glacier. The sediments settled out to form a stratified deposit.

During times of intense solar activity, the Earth's climatic temperature increased slightly, causing more glacial melting and the deposition of thicker varves. By counting the layers of thick and thin varves, scientists can establish a stratigraphic sequence that mimics both the 11-year sunspot cycle—the occurrence of numerous sunspots on the sun's surface—and the 22-year solar cycle or possibly even the early lunar cycle. The lunar cycle covers the full range of lunar orbital variations, which today is about 19 years.

THE LUNAR CYCLE

Evidence to support a solar connection with periods of drought uncovered a lunar cycle. Its effect on the climate, when combined with the solar cycle, could occasionally wreak such widespread climatic havoc as the great 1930s Dust Bowl. One of the best paleoclimate indicators is obtained from the analysis of tree rings (Fig. 111). The Moon's effect on the climate first became evident in tree rings originally analyzed for the Sun's effect on drought in the American Great Plains. The narrowing of tree rings during drought conditions revealed a tendency toward an expansion of the area of drought approximately every 22 years over the last three centuries.

Nevertheless, the influence of the 22-year solar cycle on the weather was not as significant as the 11-year cycle and accounted for only about 10 percent variance in the drought record. Therefore, the sunspots could not control drought but could only increase or decrease its likelihood. Furthermore, the effect was unpredictable, ranging from barely detectable to catastrophic.

The larger influences on the climate might be controlled to a greater extent by the Moon's 19-year cycle of changing declination, when the maximum distance from the Earth varies by about 40,000 miles. Further confirmation of the Moon's effect on the climate was made by investigating the monsoons in India. The Moon apparently accelerates or retards the seasonal march of the monsoon rains into southern Asia. The lunar cycle has also been detected in the record of floods and droughts in northern China over the last 500 years.

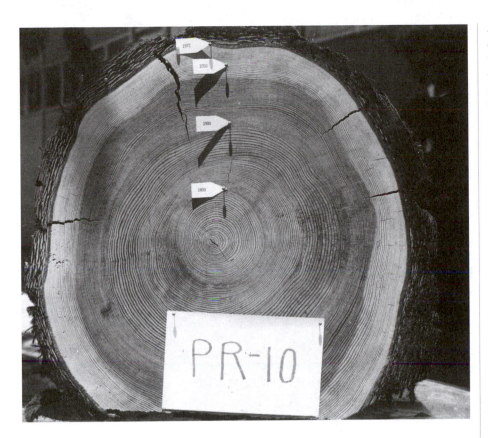

Figure 111 *A tree sample being prepared for annual growth ring studies.*

(Photo by L. E. Jackson Jr., courtesy USGS)

The Moon's influence on precipitation might result from gravity-induced tides in the atmosphere. The lunar cycle causes varying atmospheric tides just as the daily motions of the Moon raise tides in the ocean. This effect might produce a tidally induced wave in the atmosphere. Mountain ranges such as the Rockies alter the atmospheric wave, which could control weather conditions in the Plains states.

A similar atmospheric oscillation produces a wave of cloudiness high over the Indian Ocean every 40 to 50 days. As the wave sweeps eastward into the Pacific Ocean, it intensifies with speeds up to 20 miles per hour and dies out after reaching the eastern Pacific. While circling the tropics, the oscillation can set other parts of the atmosphere as far away as the polar regions pulsating at a similar frequency.

The phenomenon appears to play some role in triggering the onset and withdrawal of the monsoons in India, halting the rains in mid season. The effect could induce El Niños (Fig. 112). These generate warm currents in the equatorial Pacific that significantly affect the global climate. The atmospheric oscillation also appears to influence the upper atmospheric jet stream, which shapes the weather in North America and Eurasia.

Figure 112 *Changes in air currents during El Niño. Dashed lines are normal currents.*

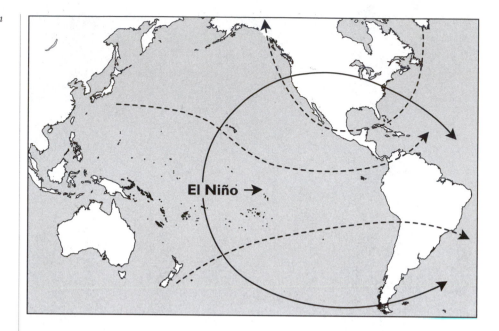

Figure 112 *Changes in air currents during El Niño. Dashed lines are normal currents.*

One of the most intriguing correlations of solar activity and the effects on the atmosphere was between the 11-year sunspot cycle and the jet stream. Data from the upper troposphere and the lower stratosphere where the jet stream flows indicate that the average position and strength of the jet stream, which influences the intensity of certain weather features, correlate strongly with solar activity. The direct solar effect on the atmosphere during times of sunspot maxima could raise the troposphere 1,500 feet, thereby affecting the jet stream, which could be useful for making long-range weather forecasts.

THE TIDAL CYCLE

Hundreds of millions of years ago, stromatolites, which are layered, moundlike structures formed by algae, recorded the interactions of the Sun, Earth, and Moon. Stromatolites found in the Bitter Springs Formation in central Australia provided an 850-million-year-old fossil record of the Sun's movement across the sky. If new sediment layers were constructed daily, then the number of layers appearing in one wavelength represented the number of days in a year during the time of the stromatolite's growth. By counting the layers in stromatolite fossils, researchers estimated that a year contained approximately 435 days during the late Proterozoic. The results agreed well with counting the growth rings of ancient coral fossils to estimate the number of days in a year going as far back as the beginning of the Cambrian period, 570 million years ago.

The studies indicated that the Earth was spinning faster on its axis and that the days were only 20 hours long in the late Precambrian. In addition, the growth patterns of ancient stromatolites contained information about the maximum travel of the Sun across the equator. Today, the Sun travels from 23.5 degrees north of the equator during the summer to 23.5 degrees south of the equator during the winter. However, about 850 million years ago, this value was about 26.5 degrees, which indicates that the climate at that time would have been much more seasonal than it is today.

Present-day stromatolites live in the intertidal zones, above the low-tide mark. Their height indicates the height of the tides, which is controlled mostly by the gravitational pull of the Moon. The stromatolite colonies of the Warrawoona group in North Pole, Western Australia are 3.5 billion years old, the oldest on Earth. They grew to tremendous heights with some more than 30 feet tall. This suggests that at an early age, the Moon was much closer to the Earth. Because of its stronger gravitational attraction at this range, it raised tremendous tides that must have flooded coastal areas several miles inland.

The data also explain why the length of day was so much shorter. The early Earth rotated much faster than it does today. As its rotational rate gradually slowed due to drag forces produced by the tides, some of its angular momentum (rotational energy) was transferred to the Moon, flinging it out into a wider orbit. Even today, the Moon is receding from the Earth at a rate of about 1.5 inches per year.

Confirmation that the Earth's spin is slowing and the Moon is moving away is found in ancient tidal fossils in a series of unusual siltstones preserved in the mountains outside Salt Lake City, Utah. The rocks called tidal rhythmites date around 800 million years old and have a pin-striped appearance caused by altering bands of dark- and light-colored material. The bands apparently formed from daily tides, which carried coarse sediments into what was once a marine estuary.

The bands thicken and thin in response to the neap and spring tide cycles, which alternate between the Moon's strongest and weakest pull on the ocean. By counting the number of neap-spring cycles per year, geologists can calculate the length of the lunar month back in time. The Utah rocks show a shorter lunar month and a day of only 21 or 22 hours long. Similar rhythmites in southeast Australia indicate that the Moon retreated only 1 inch per year 650 million years ago. This was possibly due to the shifting of the continents, which altered the shape of the ocean basins.

Ocean tides resulting from the gravitational pull of the Moon and Sun directly influence many aspects of life along the coast. The Moon revolves around the Earth in an elliptical orbit and exerts a stronger pull on the near side of the planet than on the far side. The difference between the gravitational attraction on both sides is about 13 percent, which elongates the center of gravity of the Earth–Moon system.

As the Earth spins on its axis, the oceans flow into two tidal bulges, one facing toward the Moon and the other facing away from it. The ocean is therefore shallower between the tidal bulges, giving it a slightly oval appearance. The maximum high tide in the middle of the ocean rises only about 2.5 feet. However, due to the motion of the sea and the geography of the coastline, tides often rise several times higher.

The Earth's daily rotation causes each point on the surface to go into and out of two tidal bulges. Thus, the tides appear to rise and fall twice daily. The Moon also orbits the Earth in the same direction it rotates, creeping ahead a little every day. By the time a point on the surface has rotated halfway around, the tidal bulges have moved forward with the Moon, and the point must travel farther each day to catch up. Therefore, the actual period between high tides is 12 hours 25 minutes.

The maximum tidal amplitude occurs twice monthly during the new and full Moons. This maximum occurs when the Earth, Moon, and Sun align in nearly a straight-line symmetry known as syzygy, from the Greek *syzygos* meaning "yoked together." This configuration causes spring tides, from the Saxon word *sprignam* meaning "a rising or swelling of water." The minimum tidal amplitude produces neap tides during the first and third quarters of the moon, when the Earth, Moon, and Sun align at right angles to one another and the solar and lunar tides oppose each other.

The waxing and waning of ocean tides are responsible for the prodigious growth in the intertidal zone (Fig. 113), the habitat between high and low tides. The pounding surf shapes the activity patterns of inhabitants living on beaches exposed to the open sea. Intertidal organisms occupying protected bays are not as exposed to the ocean's fluctuations. They are instead controlled by more subtle conditions such as a drop in temperature or pressure changes induced by the incoming tides, which help set the tidal rhythms.

The biologic clocks of most inhabitants of the intertidal zone are set to the rhythm of the lunar day. The rhythms are characterized by repetitive behavioral or physiological events such as feeding or resting that are synchronized to the tides. Each lunar day is about 25 hours long. It generally includes two tides with bimodal lunar-day rhythms, as compared with the unimodal solar-day rhythms of organisms attuned to the 24-hour solar day. Biologic clocks synchronized to the tidal rhythms are important survival aids by giving organisms advance warning of regular changes in the environment such as nightfall or the return of the tides. Even under constant laboratory conditions unaffected by diurnal or tidal cycles, biologic clocks continue to function, with the tidal rhythms persisting for some time.

The tidal rhythms apparently are not learned by or impressed onto organisms by the tides themselves. Crabs raised in the laboratory and exposed only to diurnal conditions exhibit a distinctive tidal component in their activ-

Figure 113 *The intertidal zone near Pillar Point, Clallam County, Washington.*

(Photo by W. O. Addicott, courtesy USGS)

ity when body temperatures are lowered. Furthermore, crabs removed from areas not subject to tides and transported to a tidal flat quickly establish a tidal rhythm. Apparently, the clock that measures the tidal frequency is innate and needs only to be activated by an outside stimulus.

Rhythmic behavior is also an expression of the genetic code. Heredity influences whether an organism is active during high or low tide. The environment plays another important part in establishing a tidal rhythm. The schedule of the tides determines only the setting of the biologic clock. Therefore, animals moved to a different ocean synchronize their clocks to the new tidal conditions.

Even when lacking outside stimuli, the biologic clock continues to run accurately but no longer controls the organism's activity. It operates independently from tidal influences until the organism is returned to the sea and the clock takes over again. Like all clocks, changes in the environment neither alter the accuracy of biologic clocks nor apply to intertidal organisms alone but to the entire spectrum of life.

THE ORBITAL CYCLES

The solar energy impinging onto the planet's surface is governed by the geometry of the Earth's orbit, the precession of the equinoxes, and the tilt of the rotational axis. Rhythmic changes in these orbital elements are called Milankovitch cycles, named for the Serbian astronomer Milutin Milankovitch, who calculated their effects on the climate in the 1920s.

The orbital cycles do not influence the total yearly insolation but alter only the amount of solar radiation reaching specific latitudes in certain seasons. A region might experience cold winters and hot summers during one cycle and mild winters and cool summers during another. Variations in the orbital motions work in concert to influence the pattern of solar radiation impinging onto the Earth. They can combine in such a manner to produce the worst possible climatic conditions.

The Earth's orbit alternates from nearly circular to elliptical about every 100,000 years, called the eccentricity cycle. When the Earth's orbit is circular, the planet maintains a constant distance of about 93 million miles from the Sun during all seasons, and insolation remains the same throughout the year. When the Earth's orbit is highly elliptical, the difference in solar input can be as much as 30 percent. The planet is closer to the Sun during one season, thereby warming it, and farther from the Sun in the opposite season, thereby cooling it. Southern Hemisphere winters would be more severe than those in the Northern Hemisphere, and northern summers would be much cooler.

The Earth is presently in an elliptical orbit. Its perihelion (closest point to the Sun) occurs in early January and its aphelion (farthest point from the Sun) occurs in early July, making southern winters slightly cooler than northern winters. The difference between perihelion and aphelion is about 3 million miles. The total insolation is 7 percent less during the northern summer than in winter.

The changing geometry of the Earth's orbit might explain the recurrence of the Pleistocene Ice Ages roughly every 100,000 years over the last 2 million years or so. Yet unexplainably, this weakest cycle, responsible for less than 1 percent of the variation of insolation, appears to produce the largest changes in climate. Perhaps it only initiates climatic changes that amplify and reinforce each other.

The shorter tilt and precession cycles alter the amount of sunlight the Northern Hemisphere receives in summer by as much as 20 percent. The tilt cycle impacts strongest on high-latitude sunlight, whereas the precession cycle has its greatest effect on sunlight in the tropics. The Earth's rotational axis presently tilts at an angle of 23.5 degrees to the plane of its orbit or ecliptic. The Sun and Moon exert a gravitational pull on the spinning Earth, causing its axis to precess or wobble like a toy top. The rotational axis thus describes a cone in space as it precesses clockwise, or in the opposite direction of the planet's rotation.

The Earth completes one precessional cycle about every 23,000 years. Therefore, around 10,000 years ago at the end of the last ice age, Vega was the

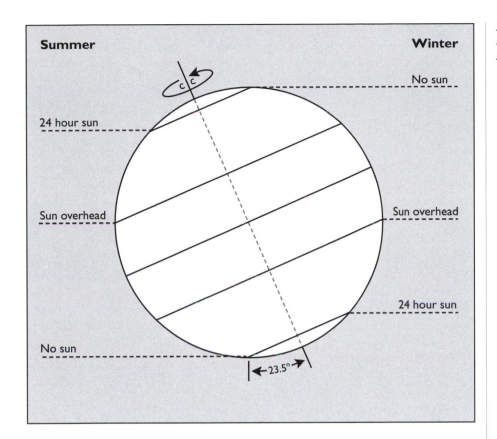

Summer   **Winter**

No sun

c c

24 hour sun

Sun overhead   Sun overhead

24 hour sun

No sun

←23.5°→

Figure 114 The effect of the Earth's tilt on the seasons.

North Star instead of Polaris today and the seasons were completely reversed. In roughly another 10,000 years, the Earth will again be tilted in the opposite direction, and winters will replace summers.

The tilt of the Earth's axis has varied from 21.5 to 24.5 degrees throughout geologic history. As the angle of tilt alters, it shifts the position of the Sun overhead during the seasons (Fig. 114). The greater the tilt angle, the larger the difference between summer and winter temperatures. If the axis were steeply inclined, the Earth would experience extremely large temperature variations from one season to the next. Conversely, if the Earth's axis did not tilt and was aligned perpendicular to the ecliptic, the planet would have no seasons.

Even slight changes in the degree of tilt can cause major climatic affects. The Earth completes one full tilt cycle about every 41,000 years. Since the end of the last ice age, the tilt angle has been steadily decreasing. This brings the prospect of cooler summers and warmer winters, ideal conditions for spawning a new ice age.

If summer sunshine in the Northern Hemisphere decreased, resulting in cooler summers, the volume of the ice sheets would grow proportionately.

The previous winter's snow would fail to melt, allowing the following winter's snow to accumulate into thick ice fields. A continuation of this condition for several years could cause the climate to revert from an interglacial to full glaciation possibly within a century.

THE HYDROLOGIC CYCLE

The Earth is the only planet in the solar system known to contain water in all three states: solid, liquid, and gas. About 2.5 percent of the Earth's water is fresh, sufficient to fill the mile-deep Mediterranean basin 10 times over. Some three-quarters of the freshwater is locked up in glacial ice, 90 percent of which buries Antarctica (Fig. 115). The remaining less than 1 percent of the freshwater is atmospheric water vapor, running water in rivers, standing water in lakes, groundwater, soil moisture, and water contained in plant and animal tissues.

At its present temperature, the atmosphere can hold only about 0.5 percent of the planet's water at any given moment. About 15 percent of the mois-

Figure 115 Cape Crozier, Antarctica showing emperor penguins.

(Photo by G.V. Graves, courtesy U.S. Navy)

172

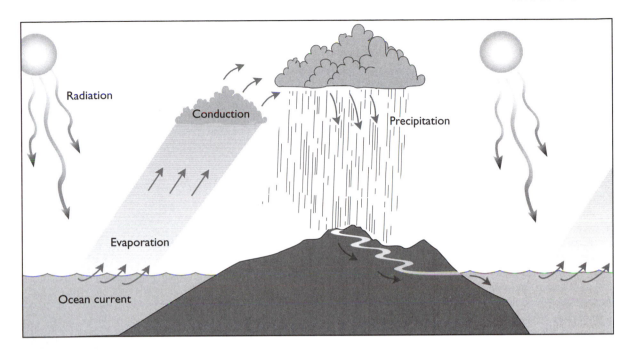

ture in the atmosphere originates from the land, with most of the rest evaporating from the ocean. If global temperatures continue to rise, more water would evaporate, increasing atmospheric water vapor, which happens to be the most effective greenhouse gas. However, the expanded cloud cover generated by the additional water vapor would probably balance out much of the greenhouse effect.

The movement of water over the Earth is one of nature's most important cycles, known as the hydrologic or water cycle (Fig. 116). Without it, life as we know it could not exist. Water takes on average about 10 days to move from the ocean to the atmosphere, cross the land, and return to the sea. The journey is only a few hours long in the tropical coastal areas and upward of 10,000 years in the polar regions. The quickest route water travels to the ocean is by stream runoff. The slowest is by glacial flow, as snow accumulating in the polar regions builds up glaciers that plunge into the sea as icebergs.

Surface runoff supplies the ocean with minerals and nutrients, and it cleanses the land of natural and human-made pollutants. The importance of water to life is so obvious it is too often overlooked. As a result, much of the surface and subsurface water has become polluted by human activities. Moreover, the accumulation of toxic substances in the ocean from runoff could irreparably damage the marine environment.

Figure 116 *The hydrologic cycle involves the flow of water from the ocean onto the land and back into the sea.*

(Courtesy USGS)

Figure 119 *The biologic carbon cycle. It converts atmospheric carbon dioxide directly into vegetation or animal matter, which reverts back to carbon dioxide upon decay or combustion.*

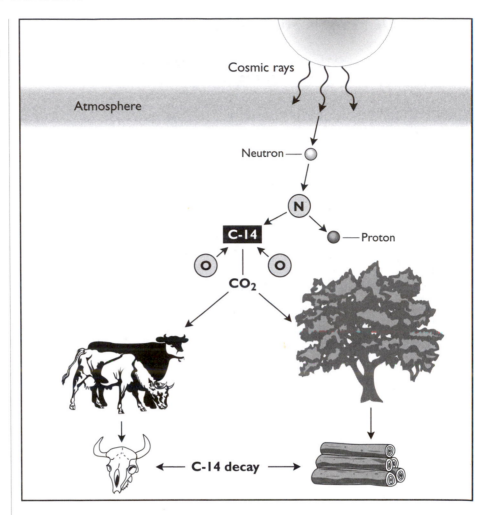

Atmospheric carbon dioxide therefore operates like a thermostat to regulate the temperature of the planet. Even slight changes in the carbon cycle would have considerable influence on the climate. The Earth cools if the carbon cycle removes too much carbon dioxide and warms if the carbon cycle generates too much carbon dioxide.

The creation and decomposition of peat bogs might have been responsible for most of the changes in levels of atmospheric carbon dioxide during the past two glaciations. The bogs have accumulated upward of 250 billion tons of carbon in the last 10,000 years since the end of the last ice age, mostly in the temperate zone of the Northern Hemisphere. Over geologic time, progressively more land has drifted into latitudes where large quantities of carbon are stored as peat. During the last million years, glaciations have gradually remolded large parts of the Northern Hemisphere into landforms more suitable for peat bog formation in wetlands (Fig. 120)

The oceans have the largest influence on atmospheric carbon dioxide levels. The mixed layer of the ocean, within the upper 250 feet, contains as much carbon dioxide as the entire atmosphere. The concentration of gases in the upper layers of the ocean always maintains equilibrium with the atmosphere. The gas dissolves into seawater mainly by the agitation of surface waves. If the ocean were lifeless, bereft of photosynthetic organisms to absorb the dissolved carbon dioxide, much of its reservoir of this gas would escape into the atmosphere, more than tripling the present content. This would result in a runaway greenhouse effect.

Atmospheric carbon dioxide combines with rainwater to form carbonic acid. This chemically reacts with surface rocks, yielding dissolved calcium and bicarbonate. These substances are carried by streams to the sea, where marine organisms incorporate them into their bodies to build calcium carbonate skeletons and other supporting structures. When the organisms die, their skeletons sink to the ocean bottom and are dissolved in the cold, deep waters of the abyss generally at depths greater than two miles. Because of its large volume, the abyss holds the largest reservoir of carbon dioxide, containing 65 times more carbon than the entire atmosphere.

Most carbon accumulates into sediments on the ocean floor and on the continents. In shallow water, skeletons from once-living organisms build thick

Figure 120 *Wetland area in Doge County, Wisconsin.*

(Photo by Ron Nichols, courtesy USDA Soil Conservation Service)

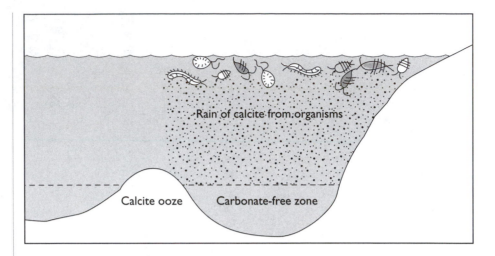

deposits of carbonate rock, such as limestone (Fig. 121), storing massive amounts of carbon in the geologic column. The burial of carbonate in this manner is responsible for about 80 percent of the carbon deposited onto the ocean floor. The rest originates from the burial of dead organic matter washed off the continents.

In this respect, marine life acts as a pump to remove carbon dioxide from the atmosphere and the ocean's surface waters and to store it in the deep sea. The faster this biologic pump works, the more carbon dioxide that is removed from the atmosphere. The rate is determined by the amount of nutrients in the ocean. A reduction of nutrients slows the biologic pump, returning deep-sea carbon dioxide to the atmosphere.

Half the carbonate transforms back into carbon dioxide, which returns to the atmosphere, mostly by upwelling currents in the tropics. This explains why the concentration of atmospheric carbon dioxide is highest near the equator. Without this process, in a mere 10,000 years, all carbon dioxide would be taken out of the atmosphere. The loss of this important greenhouse gas would result in the cessation of photosynthesis and the extinction of life.

Volcanic activity on the ocean floor and on the continents is the final stage of the carbon cycle. Volcanoes return carbon dioxide to the atmosphere (Fig. 122). The carbon dioxide escapes from carbonaceous sediments as they melt in the Earth's interior and become part of the magma. The molten magma along with its content of carbon dioxide rises to the surface to feed magma chambers beneath volcanoes. When the volcanoes erupt, carbon dioxide and other volatiles are released from the magma, sometimes violently, and enter the atmosphere.

Volcanoes also expel significant quantities of nitrogen compounds. Atmospheric nitrogen originated from early volcanic eruptions and the break-down of ammonia, a molecule of one nitrogen atom and three hydrogen atoms and a large constituent of the primordial atmosphere. Unlike most other gases, which have been replaced or permanently stored in the crust, the Earth retains much of its original nitrogen. Nitrogen comprises 79 percent of the atmosphere and is a major constituent of life. Carbon, hydrogen, and nitrogen are the essential elements for manufacturing proteins and other biologic molecules. Nitrogen is practically an inert gas and requires special chemical reactions to be used by nature. Therefore, much energy is required to make nitrogen combine with other substances.

The nitrogen cycle is a continuous exchange of elements between the atmosphere and biosphere by the action of organisms, such as nitrogen-fixing bacteria. All methods of nitrogen fixation need a source of abundant energy, mainly supplied by the Sun. The Earth also provides a source of energy used by animals living near hydrothermal vents on the deep ocean floor. The decay of organisms after death releases nitrogen back into the atmosphere, thus forming a closed cycle. Life prevents all nitrogen from transforming into nitrate, which easily dissolves in the ocean, where denitrifying bacteria return the nitrate–nitrogen to its original gaseous state. Without this process, all nitro-

Figure 122 *Volcanoes contribute large amounts of water vapor, carbon dioxide, and other important gases to the atmosphere as well as contribute to the growth of continents.*

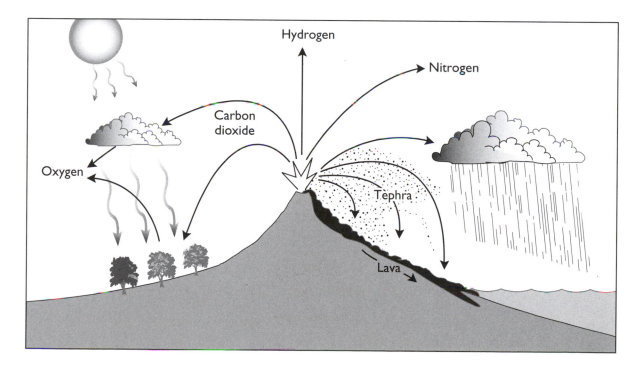

gen in the atmosphere would have long ago disappeared, and the Earth would be left with only a fraction of its current atmospheric pressure.

This ends the discussion on how life has been shaped down through geologic history as depicted in the previous chapters. The next chapter will take a journey back in time in search of some of the strangest creatures that have ever lived on Earth.

9

UNIQUE LIFE-FORMS
THE MOST UNUSUAL ADAPTATIONS

This chapter will examine some of the strangest creatures to be shaped by evolution in historical geology. During its long history, the Earth underwent many environmental changes. Species evolving to meet these challenging conditions were molded by adaptation into some of the most bizarre creatures ever discovered. Many evolutionary pathways weave along branches of the tree of life. Species leave their footprints in the fossil record, itself only a modest representation of all those that have gone before.

Practically every conceivable biologic form and function have been tried, some more successfully than others. Through this trial-and-error method of speciation, natural selection has selected certain species to prosper, while others have become evolutionary dead ends. Most species have been discarded by extinction, never to be seen again, providing today's world with a myriad of unique life-forms.

THE ENIGMATIC EDIACARANS

The Ediacara Formation in South Australia's Ediacara Hills contains fossil impressions of highly curious creatures (Fig. 123). Some of them were giants

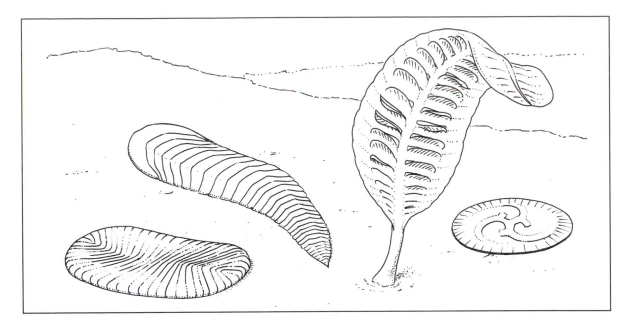

Figure 123 *The late Precambrian Ediacara fauna.*

in their day, reaching 3 feet in size and ranging in shape from spoked wheels and miniature anchors to corrugated and lettuce-like fronds. The Ediacaran behemoths provide the first evidence of complex life on the planet, which was previously populated by microscopic, single-celled organisms. The complex life-forms abruptly came into existence around 600 million years ago, not long after the last major episode of late Precambrian glaciation, possibly the most severe in Earth history. The emergence of the Ediacaran fauna was closely linked to profound changes in the environment, including the breakup of continents and the increased oxygen that made possible the evolution of large animals.

More than 30 species of these simple, beautiful creatures have been identified from fossil impressions in rock from 20 sites scattered around the world. Many impressions were left by animals apparently related to modern jellyfish, corals, and segmented worms. Others appear to represent arthropods, annelids, and possibly even echinoderms. The bodies of some forms had threefold symmetry unlike any seen in modern organisms.

The fossils represent a marine life very different from that of today's oceans, including feathery fronds, puckered pouches, flattened blobs, and engraved disks. Many were marked with radiating, concentric, or parallel creases, whereas others were inscribed with delicate branches. The organisms seemed to have no heads or tails; circulatory, nervous, or digestive systems; eyes, mouths, bones, muscles; or internal organs, making classification very difficult.

The Ediacaran fauna are found worldwide in rocks that just predate the Cambrian explosion, when shelly faunas first erupted onto the scene. Indeed, the disappearance of the Ediacarans made the ensuing Cambrian explosion possible. Only when the Precambrian seas were cleared of primitive life-forms could more advanced species flourish and diversify. The evolutionary heyday that marked the beginning of the Cambrian gave birth to most animal phyla that presently swim, crawl, or fly around the globe. Animals appeared for the first time covered with shells, bearing jaws, claws, and other biologic innovations.

The Ediacaran organisms appeared to be a preview of these animals and possessed body styles or morphologies never seen before or since in the fossil record. Most, if not all, were not related to modern forms. They were perhaps a failed evolutionary experiment in multicellular organisms completely separate from all known kingdoms of life and were wiped out in a previously unrecognized mass extinction. Yet some Ediacaran organisms apparently inhabited the planet far longer than previously thought, even surviving well into the Cambrian.

The fossils displayed an unusual body architecture totally alien to anything seen on Earth today. Some Ediacaran fauna called vendobionts, after the Vendian period, the final stage of the Precambrian, evolved a unique solution to the problem of growing large bodies. They employed networks of tubes, such as blood vessels, to transport nutrients and oxygen to individual cells. The fossils show no openings for ingesting food and eliminating wastes and probably absorbed oxygen and nutrients directly from seawater or harbored symbionic algae that converted sunlight into energy. They showed no obvious internal digestive or circulatory systems, which apparently did not fossilize. Yet some animals did leave what appear to be fecal pellets, indicating an advanced digestive system.

The extremely flattened bodies of the Ediacaran fauna maximized the ratio of surface area to volume, enabling the efficient absorption of nutrients and oxygen and the collection of light for symbiotic algae. The algae lived embedded within the tissues of the host animals, which offered protection from predation in return for nutrients and the removal of waste products. These adaptations served well for the prevailing marine conditions of the late Precambrian, when shallow seas were poorly supplied with nutrients and oxygen.

The increasing oxygen content advanced the evolution of large animals. Many profound physical changes occurred, prompting a rapid radiation of Ediacaran fauna. The increased marine habitat area during continental breakup spurred the greatest diversity of new species the world has ever known, when seas contained large populations of widespread and unusual organisms.

Figure 124 *Helicoplacus was an experimental species, which became extinct about 510 million years ago, whose body parts were arranged in a manner not found in any living creature.*

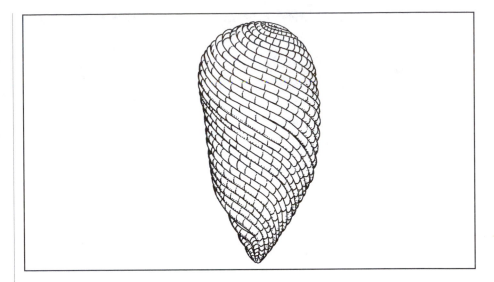

Figure 124 *Helicoplacus was an experimental species, which became extinct about 510 million years ago, whose body parts were arranged in a manner not found in any living creature.*

The Ediacaran fauna spawned from adaptations to highly unstable conditions during the late Precambrian. Overspecialization to a narrow range of environmental conditions, however, brought about a major extinction of Ediacaran species around 540 million years ago at the very doorstep of the Cambrian explosion. Marine animals that survived the die-off were markedly different from their Ediacaran ancestors and participated in the greatest diversification of new species the world has ever known.

Never had so many experimental organisms existed, none of which have any modern counterparts. One example is helicoplacus (Fig. 124). Its body parts were constructed in a manner not found in any living organisms. It was about 2 inches long and shaped like a spindle covered with a spiraling system of armor plates. It emerged during the transition from Precambrian to Cambrian, when more unusual body plans arose than at any other time in Earth history. As with most species of the early Cambrian, helicoplacus did not survive long term and became extinct about 510 million years ago, after only a comparatively brief 20-million-year existence.

BIZARRE BURGESS SHALE FAUNAS

Early in the 20th century, the remains of exceptional soft-bodied and shelly animals were discovered in the Burgess Shale Formation of British Columbia, Canada. The Burgess Shale faunas first appeared in the lower Cambrian about 540 million years ago soon after the emergence of complex organisms. The assemblage featured more diversity of basic anatomic designs than in all the world's oceans today. It included some two dozen types of plants and animals that have no modern counterparts and are thoroughly mysterious. Indeed, humans owe our own existence to a leechlike animal called pikaia (Fig. 125). This was the first known member of the chordate phylum and one of the rarest Burgess Shale fossils.

The Burgess Shale faunas were surprisingly complex with specialized adaptations for living in a variety of environments. Some species appear to be surviving Ediacaran fauna, most of which became extinct near the end of the Precambrian. Indeed, the Cambrian explosion might have been triggered in part by the availability of habitats vacated by departing Ediacaran species. The Burgess Shale faunas comprised more than 20 distinct body plans. Although many mass extinctions and proliferations of marine organisms have occurred since, no fundamentally new body styles have appeared during the past 500 million years.

The ultimate fate of the Burgess Shale faunas is unclear. Most species appear to have suffered an abrupt extinction in the late Cambrian. Only a few organisms evolved into anything living today. Many of these peculiar animals were possibly carried over from the upper Precambrian but never made it beyond the middle Paleozoic. They were so strange, they defied efforts to classify them into existing taxonomic groups.

The Burgess Shale faunas originated in shallow water on a gigantic coral reef that dwarfed Australia's Great Barrier Reef, the single most massive structure built by present-day living beings. The ancient reef surrounded Laurentia, the ancestral North American continent. The reef was covered with mud that readily trapped and fossilized organisms. Most occurrences originated from the western Cordillera of North America, an ancient mountain range that faced an open ocean in the middle Cambrian. Similar

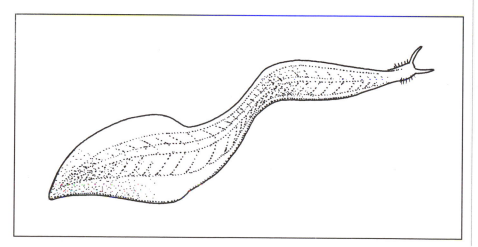

Figure 125 *Pikaia was one of the earliest chordates.*

faunas existed on other cratons, including the North and South China blocks, Australia, and the East European platform. Their widespread distribution around other continents suggests many members had a swimming mode of life.

Most Burgess Shale faunas abruptly went extinct at the end of the Cambrian. Only a few archaic forms survived to the middle Devonian. Had they prevailed and produced descendants, our planet would now be graced by an entirely different set of life-forms. One peculiar animal aptly named hallucigenia (Fig. 126), to honor its strangeness, was a wormlike creature that propelled itself across the seafloor on seven pairs of legs. Two rows of spines were used for protection.

A curious Burgess Shale animal called wiwaxia (Fig. 127) was a spiny creature about an inch long, possibly related to the modern scaleworm known as a sea mouse. It resembled an undersea porcupine, with large scales and two rows of spikes running along its back apparently for thwarting predators. It fed by scraping off fragments of food with a rasping organ that resembled a horny, toothed tongue.

An unusual Burgess Shale worm had enormous eyes and prominent fins. A chunky burrowing carnivorous worm called ottoia was found with shelly animals still in its gut. When reaching out from the bottom mud, it extended a muscular, toothed mouth and swallowed its prey whole. An odd creature called opabinia (Fig. 128) had five eyes arranged across its head, a vertical tail fin to help steer it along the seafloor, and a grasping organ projected forward possibly used for catching prey. Its long, hoselike front appendage bestowed upon it the name "swimming vacuum cleaner."

Figure 126 Hallucigenia is one of the strangest animals preserved in the fossil record.

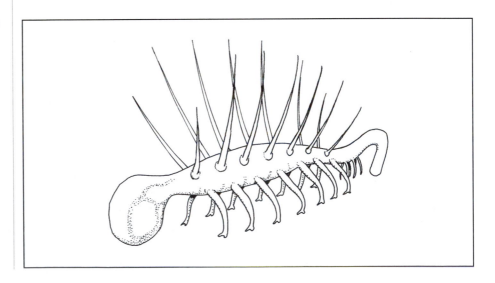

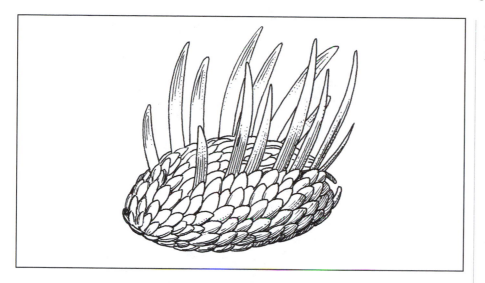

Figure 127 Wiwaxia was well protected against predators.

About 40 percent of the Burgess Shale faunas consisted of arthropods, with fossil remains of about 20 different arthropod possibilities that did not survive. One giant arthropod found in the middle Cambrian Burgess Shale formation of western Canada was as much as 3 feet long. The most interesting of all Burgess Shale arthropods is aysheaia (Fig. 129), an animal with a pudgy body and stubby limbs. An extraordinary arthropod called anomalocaris (Fig. 130), whose name means "odd shrimp," was possibly the largest of the Cambrian predators as well as the oldest-known sizable predator in the fossil

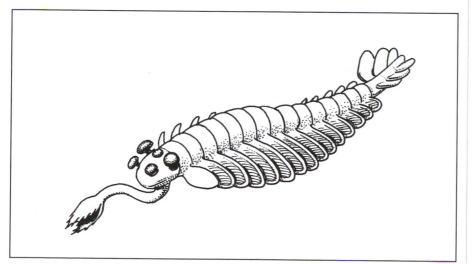

Figure 128 Opabinia had a forward-projecting grasping organ for catching prey.

Figure 129 *Aysheaia was an early Cambrian, stubby-legged animal.*

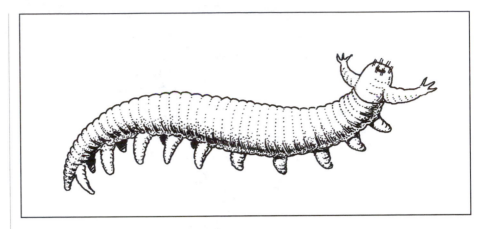

record. This earliest of monsters reached 6 feet in length and had a hideous mouth surrounded by spiked plates on the underside of the body. Up to eight rings of teeth reaching 10 inches across were arranged concentrically, leading into the mouth.

Anomalocaris propelled itself by raising and lowering a set of side flaps, swimming in a manner similar to modern manta rays. It also had a broad tail with a pair of long, trailing spines used for steering and stability. The body was flanked by a pair of jointed appendages apparently designed for holding and crushing the armored plates of invertebrates. The animal appeared to be well equipped for devouring crustaceans and is appropriately dubbed the "terror of trilobites." Many trilobite fossils are found with rounded chunks

Figure 130
Anomalocaris was a fierce predator of trilobites.

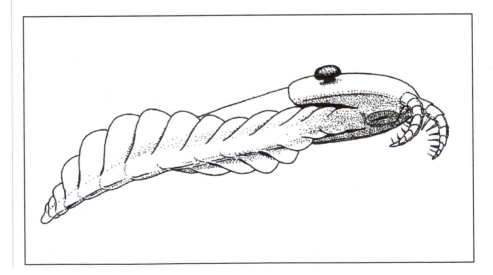

bitten out of their sides, signifying they had somehow escaped the terrifying grasp of anomalocaris.

LILIES OF THE SEA

The echinoderms are perhaps the strangest animals preserved in the fossil record and are among the most prolific marine species. They are remarkable animals with fivefold and bilateral symmetries and exoskeletons composed of numerous calcite plates. The most successful echinoderms were the crinoids (Fig. 131). They were commonly called "sea lilies" because of their resemblance to flowers supported by long stalks anchored to the seabed. The earliest known crinoid in the fossil record was a Burgess Shale echinoderm that displayed a number of primitive features.

The crinoids were the dominant echinoderms of the middle and upper Paleozoic, with many species still prospering today. Some crinoids were also free-swimming types. The modern floating-swimming crinoids underwent a short, widespread evolutionary burst in the late Cretaceous and are useful for dating rocks of that time.

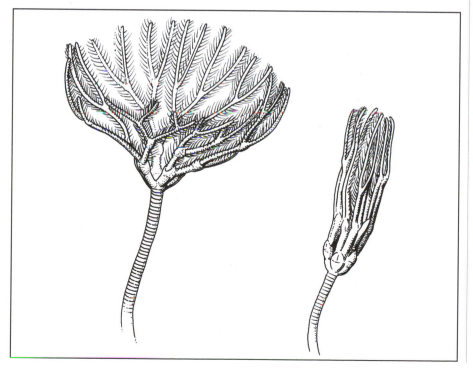

Figure 131 Crinoids grew upward of 10 feet or more tall.

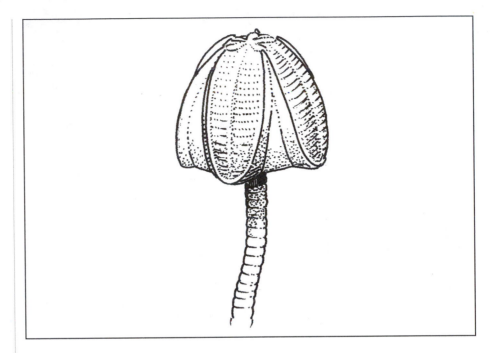

The stalks of crinoids reached 10 feet or more in length. However, during the exuberant growing conditions of the warm Mesozoic seas, they grew upward of 60 feet long. The stalks comprised perhaps a hundred or more calcite disks called columnals and were attached to the ocean floor by a rootlike appendage. Star-shaped columnals were common in the Triassic and Jurassic periods. The crinoid stalks are especially abundant as fossils. On weathered limestone outcrops, the stalks often appear as long strings of beads. Crinoidal limestones are sedimentary rocks composed almost entirely of crinoid parts, especially the stem plates.

A cup called a calyx perched atop the stalk housed the digestive and reproductive organs. The animal strained food particles from passing water currents with five feathery arms that extended from the calyx, giving the crinoid its typical flowerlike appearance. The extinct Paleozoic crinoids made excellent fossils as well as their blastoid relatives (Fig. 132), whose calyxes resembled rose buds comprising regularly arranged plates. The blastoids became extinct at the end of the Paleozoic due to stiff competition from the crinoids for scarce shallow-water platform habitats. The ancient crinoid and blastoid fossils are eagerly sought after by amateur fossil hunters.

JET-PROPELLED MARINERS

Cephalopods have a unique form of transportation. They traveled by jet propulsion, sucking water into a cylindrical cavity from openings on each side of the head and expelling it under pressure through a funnellike appendage. The nautiloids with their straight, streamline shells up to 30 feet and more in length were among the swiftest animals of the ancient seas. Sometimes, several fossil shells are found one inside another like a stack of ice cream cones, possibly being driven together by undersea currents.

The ammonites possessed a large variety of coiled shell forms and were the most spectacular of marine predators. Their external shells were subdivided into air chambers. The suture lines joining the segments presented a variety of patterns used for identifying various species. The air chambers provided buoyancy to counterbalance the weight of the growing shell. Most shells were coiled in a vertical plane, some forms were spirally coiled, and others were essentially straight, which often made swimming awkward. Their large variety of coiled shell forms made the ammonites ideal for dating Paleozoic and Mesozoic rocks. Throughout the Mesozoic, ammonite shell designs steadily improved. The cephalopod became one of the swiftest creatures of the deep, competing successfully with fish.

For 350 million years, 10,000 ammonite species roamed the seas. Of the 25 families of widely ranging ammonites living in the late Triassic, all but one or two became extinct at the end of the period, when half of all species died out. The ammonites that managed to escape extinction eventually evolved into scores of ammonite families during the Jurassic and Cretaceous. The ammonites lived mainly at the middle depths and might have shared many features with living squids and cuttlefish. Some ammonites grew to tremendous size with coiled shells up to 7 feet wide and straight shells 20 feet or longer. The nautilus, commonly referred to as a "living fossil" because it is the ammonite's only living relative, lives at extreme depths of 2,000 feet.

The belemnoids, with long, bulletlike shells, originated from more primitive nautiloids and were related to the modern squid and octopus. They were abundant during the Jurassic and Cretaceous and became extinct by the Tertiary. The shell was straight in most species and loosely coiled in others. The chambered part of the shell was smaller than that of the ammonite, and the outer walls thickened into a fat cigar shape.

After the extinction that killed off the dinosaurs and large numbers of other species, all shelled cephalopods disappeared except the nautilus, found exclusively in the deep waters of the South Pacific and Indian Oceans. Other surviving cephalopods included shell-less species such as the cuttlefish, octopus, and squid (Fig. 133). The octopus is somewhat like an alien life-form from

Figure 133 *The squids were among the most successful cephalopods.*

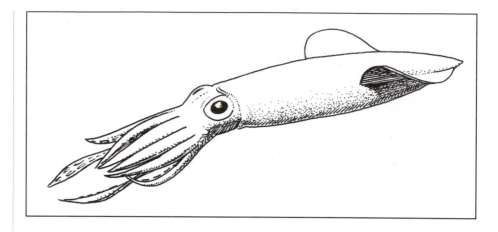

another world, for it is the only species whose blood is based on copper instead of iron as with all other earthly creatures.

SEAFARING REPTILES

Reptiles are among the most diverse group of animals, having occupied land, sea, and air. Reptiles returned to the sea during the Mesozoic to compete with the fish for a plentiful food supply. They included the sea serpentlike plesiosaurs, the sea cow–like placodonts, and the dolphinlike mixosaurs (Fig. 134). The 150-million-year-old pachycostasaurs, meaning "thick-ribbed lizard," were carnivorous marine reptiles 9 feet or more long. They resembled plesiosaurs, with thick, heavy ribs possibly used to house very large lungs and for ballast in order to hunt at great depths.

Figure 134 *The mixosaur was a dolphinlike reptile that returned to the sea.*

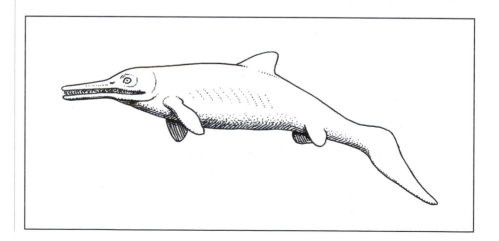

The sharklike ichthyosaurs, whose name means "fish lizard," were reptiles that returned to the sea to compete with fish where they achieved great success. They were fast-swimming, shell-crushing marine predators that apparently preyed on ammonites. The animal would first puncture the shell from the victim's blind side, causing it to fill with water and sink to the bottom, where the attack could be made head on. Puncture marks on fossil ammonite shells are spaced the same distance apart as fossil ichthyosaur teeth, suggesting these highly aggressive predators might have hastened the extinction of most ammonite species by the end of the Mesozoic.

The placodonts were a group of short, stout, marine reptiles with large, flattened teeth. They probably fed primarily on bivalves and other mollusks. Several other reptilian species also went to sea, including lizards and turtles that were quite primitive. Many modern giant turtles are descendants of those marine reptiles. However, only the smallest turtles survived the extinction at the end of the Cretaceous. Turtles are the closest living relatives of crocodiles, another extremely successful aquatic reptile. The crocodiles, which passed through the extinction practically unscathed, are closer to turtles than they are to lizards, snakes, and birds. Turtles, which are holdovers from an ancient group called anapsids that lacked holes in the sides of their skulls, were once regarded as outsiders among modern reptiles. This is because living reptiles and birds termed diapsids have two holes in the sides of their skulls.

One of the strangest reptiles was tanystropheus (Fig. 135), dubbed the "giraffe-neck saurian." The animal measured as much as 15 feet from head to

Figure 135
Tanystropheus had an extremely long neck, more than twice the length of its trunk.

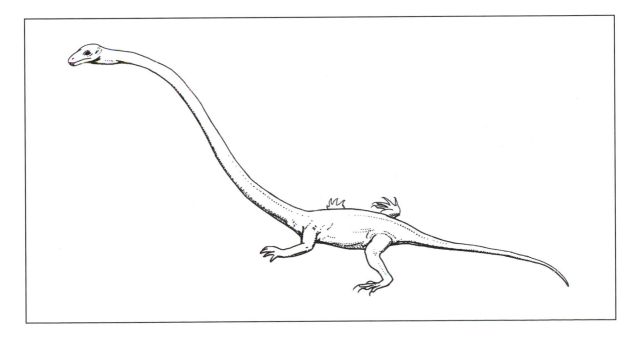

tail and is famous for an absurdly long neck, more than twice the length of the trunk. As the reptile grew, the neck outpaced the rest of its body. Apparently, the tanystropheus was aquatic because on land it could not possibly have supported the weight of such a grossly elongated neck, which probably evolved for scavenging bottom sediments for food.

THE GREATEST AVIATORS

The flying reptiles called pterosaurs, including the ferocious pterodactyls, originated in the early Jurassic. With wingspans up to 30 feet or more long, they were by far the largest animals ever to fly, dominating the skies for more than 120 million years. The fourth finger of each forelimb was extremely elongated and supported the front edge of a membrane that stretched from the flank of the body to the finger tip, leaving the other fingers free for such purposes as climbing trees. A bat's wing, by comparison, is constructed by lengthening and splaying all fingers, which are covered with membrane.

The great pterosaur debate began to flare up in 1817 when the German naturalist Thomas von Soemmerring reconstructed a fossil skeleton believed to have belonged to a young pterodactyl. He portrayed the animal as an ungainly glider with leathery wings attached on each side of the body down to the legs and stretched between a grotesquely elongated finger. Such a large flap of skin would have made the creature a clumsy walker, shuffling around on all fours like a bat, not a good way to take off. Later reconstructions showed a flying reptile with thin birdlike wings attached to the hip, allowing the animal to walk on two legs, a better design for gaining speed for liftoff.

Why pterosaurs took to the air in the first place remains a mystery. They apparently arose from tree-dwelling reptiles rather than from running ground dwellers. Their ancestors might have grown skin flaps for jumping from tree to tree in a manner similar to flying squirrels. The wing membranes might have originally served as a cooling mechanism that regulated body temperature by fanning the forelimbs, and through natural selection they eventually became flying appendages.

In their overall structure and proportions, pterosaurs resembled both birds and bats, with the smallest species measuring roughly the size of a sparrow. Like birds, their hollow bones conserved weight to make flight possible. The larger pterosaurs were similarly proportioned to a modern-day hang glider and weighed about as much as the pilot. Many pterosaurs had tall crests on their skulls, possibly functioning as a forward rudder to steer the reptiles in flight.

Most pterosaurs living in North America about 75 million years ago were thought to be toothless, until the first evidence of a flying reptile with teeth assumed to be an inch long was discovered in Texas. The fossil record of pterosaurs is meager because their delicate bones were easily destroyed. The only remains found of the Texas specimen was its bill, which had a distinctive crest that apparently helped stabilize the pterosaur's head like a keel as it dipped into the sea to catch fish.

The discovery of the 100-million-year-old Texas pterosaur proves that teeth-bearing flying reptiles similar to those in South America and Eurasia migrated to North America. The pterosaur had a wingspan of about 5 feet and a slender body about 1.5 feet long. It resembled a species that lived in Great Britain about 140 million years ago. Whether any connection exists between this toothy North American pterosaur and the toothless variety that came later is still unclear.

The largest creature to take to the air was a pterosaur called *Quetzalcoatlus northropi*. It was about the size of a small airplane and had a wingspan of 30 feet. Fossils of the huge flier were unearthed in the Big Ben National Park in southwest Texas. Although originally thought to feed on carrion such as dead dinosaurs, it was most likely a fish eater resembling a giant pelican even though the deposits where the animal was found were far from the sea. Their extremely large size enabled them to fly long distances to the ocean, where they could find plentiful fish on which to prey.

Pterosaurs might have flown by jumping off cliffs and riding the updrafts, by climbing trees and diving into the wind, or by gliding across the tops of wave crests like modern albatrosses. The animal might have taken off by trotting along the ground flapping its wings the way a gooney bird does to gain speed. Alternatively, it might have simply stood on its hind legs, caught a strong breeze, and took to the air with a single flap of its huge wings and a kick of its powerful legs. It probably spent most of its time aloft riding air currents as with present-day condors.

When landing, it simply stalled near the ground, gently touching down on its hind legs in a similar manner to how a hang glider lands. While on the ground, pterosaurs might have been ungainly walkers, sprawling about on all fours like a bat. The structure of the ankle, knee, and hip joints precludes pterosaurs from flying as bats do, however, because they could not have brought their legs up in the same plane as the wings. Fossil pterosaur pelvises suggest that the hind legs extended straight down from the body, enabling the reptiles to walk upright. They could then trot along for short bursts to gather speed for takeoff. That the pterosaurs could really fly leaves little doubt. They went on to become the greatest aviators the world has ever known.

BROODING DINOSAURS

Dinosaurs are perhaps the most celebrated species ever to have sprouted from the family tree of animals. During their 150-million-year reign over all other terrestrial creatures, every dinosaur species could trace its origin back to a single common ancestor called eoraptor, meaning "early hunter." It evolved about 240 million years ago from predatory reptiles. Early in dinosaur evolution, the animals had to compete with huge crocodiles, flying reptiles called pterosaurs, and other fierce reptiles. However, by the end of the Jurassic, dinosaurs were the largest predators ever to walk the Earth.

More than 500 dinosaur genera have been discovered over the last two centuries. Dinosaurs are often portrayed as lumbering, unintelligent brutes. Yet the fossil record suggests that many species, such as velociraptor, meaning "speedy hunter," were swift and intelligent. Nor were all dinosaurs giants, and many species were no larger than most mammals today. The protoceritops, a parrot-beaked, shield-headed dinosaur, and the ankylosaur, whose name means "stiff lizard," an elephant-sized dinosaur heavily covered with protective bone and swinging a club-shaped tail, were very common. They ranged over wide spaces like modern-day sheep.

Perhaps no other dinosaur has captured the imagination more than *Tyrannosaurus rex*. It has reigned as the most fearsome and favorite dinosaur, a tyrant unparalleled by carnivores on any continent. However *T. rex* was at a distinct disadvantage as a galloping predator. It was bipedaled with very strong legs but with weak forelimbs unable to break a fall. If a 7-ton *T. rex* charging prey at a top speed of 25 miles per hour were to trip and fall, the crash could be fatal. More likely, the dinosaurs traveled in hunting packs that surrounded their prey, requiring much less speed and thereby resulting in fewer crashes.

The smaller dinosaurs possessed hollow bones similar to those of birds. Indeed, birds are the only living relatives of dinosaurs. The skeletons of many small dinosaurs closely resembled those of birds, suggesting they descended from dinosaurs. Some dinosaurs had long, slender hind legs, long delicate forelimbs, and a long neck. If not for a lengthy tail, the skeletons would resemble those of modern ostriches. Certain feathered dinosaurs might have been the progenitors of birds. The feathers would also provide thermal insulation during a colder climate.

Some dinosaur species might have acquired a degree of temperature control similar to mammals and birds. Dinosaurs descended from the thecodonts, the same common ancestors of birds, the dinosaur's distant living relatives. An argument in favor of warm-blooded dinosaurs contends that the skeletons of smaller, lighter species bear many resemblances to those of warm-blooded birds. Evidence for rapid juvenile growth, which commonly occurs among mammals, is also exhibited in the bones of some dinosaur species, providing another sign of warm-bloodedness.

Warm-blooded animals grow rapidly until reaching maturity, whereas cold-blooded ones continue to grow steadily until death. A comparison among the bones of dinosaurs, crocodiles, and birds, all of which apparently had a common ancestor, show a similarity between bird and dinosaur bones, a possible warm-blooded connection. A comparison of the density of pelvic bones in fossilized dinosaur embryos with the bones of modern crocodiles and birds indicated that juvenile dinosaurs were highly mobile, requiring the energy only a warm-blooded body could supply.

Furthermore, the blood vessel density of dinosaur bones was higher than living mammals, another warm-blooded hallmark. Some dinosaur skulls display signs of sinus membranes, which exist only in warm-blooded animals. Respiratory turbinates, which act as thermal exchangers common in the nostrils of warm-blooded species, appear to have been present in dinosaurs as well. Bones of carnivorous dinosaurs found on Antarctica suggest that they either adapted to the cold and dark by being warm-blooded or migrated to nearby South America across a land bridge.

Analysis of dinosaur tracks suggests that some species were swift and agile, requiring a high rate of metabolism that only an endothermic body can provide. The anatomy of some dinosaurs implies they were active, upright, and constantly feeding, a behavior that suggests a high metabolic rate commensurate with warm-blooded animals. They breathed with diaphragm-driven lungs for the extra respiration needed for rapid movement, much like modern mammals. However, when at rest, they might have switched to rib-based respiration similar to reptiles, giving them a metabolism functioning unlike that of any living animal.

Measurements of trackways can determine the stride of the animal and sometimes even its pace, whether walking or running. Dinosaur tracks are the most impressive of all fossil footprints (Fig. 136) because the great weight of many species produced deep indentations in the ground. Their footprints exist in relative abundance in terrestrial sediments of the Mesozoic age throughout most of the world. The study of dinosaur tracks suggests that some species were highly gregarious and gathered in herds. Large carnivores such as *Tyrannosaurus rex* were swift, agile predators that could sprint at speeds up to 20 miles per hour or more according to their tracks.

The giant herbivorous dinosaurs might also have traveled in large herds, with the adults in the lead and the juveniles kept in the center for protection. The duck-billed hadrasaurs were among the most successful of all dinosaur groups. They grew to 15 feet tall and lived in the Arctic regions, where they either adapted to the cold and dark or migrated en masse over long distances to warmer climates. The complex social behavior of dinosaurs appears to be

Figure 136 *Dinosaur tracks, Navajo Canyon, Coconino County, Arizona.*

(Photo by H. E. Gregory, courtesy USGS)

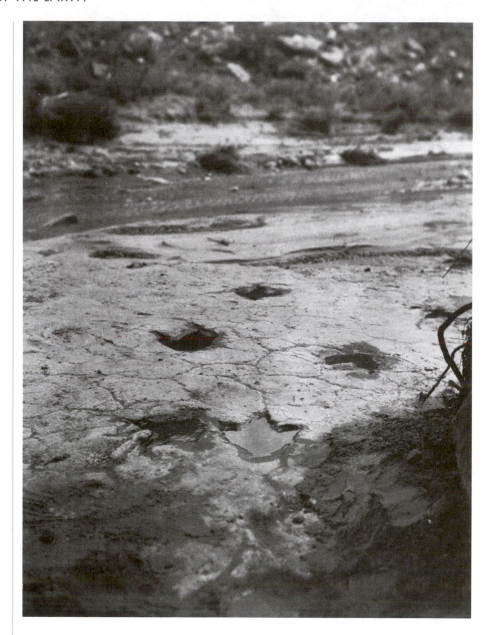

an evolutionary advancement that resulted from having a warm body temperature. Even the females of some dinosaur species might have had live births like mammals. Other dinosaurs appear to have made nests, brooded, and tenderly guarded their unhatched eggs as birds do.

The newborn dinosaurs were apparently not defenseless dependents as with most birds but leaped from their shells ready to run and protect themselves from a dangerous world. The infant dinosaurs had the bone and muscle strength to move rapidly and scamper away from danger the way modern crocodile young do. However, a fossil discovery of maiasaura, which literally means "good mother lizard," suggests that dinosaur adults were tender, nurturing parents whose young remained in their nest for feeding and protection.

Females of some dinosaur species nurtured and fiercely protected their young until they could fend for themselves. This allowed larger numbers to mature into adulthood, thereby ensuring the continuation of the species. Furthermore, dinosaurs had a long life span, and fewer juveniles were needed to sustain an adult population of a given size. The parents might have brought food to their offspring and regurgitated seeds and berries as is the custom of modern birds. This parental care for infants suggests strong social bonds, which help explain why the dinosaurs remained so successful.

Much information can be obtained about dinosaur diets by studying coprolites. These are masses of fecal matter preserved as fossils and are usually modular, tubular, or pellet shaped. Dinosaur coprolites can be quite massive, even larger than a loaf of bread. Coprolites are often used to determine the feeding habits of extinct animals. For example, coprolites of herbivorous dinosaurs are black, blocky shaped, and usually filled with plant material. Those of carnivorous dinosaurs are spindle shaped and contain broken bits of bone from dining on other animals. Some dinosaur species swallowed gizzard stones similar to the way modern birds do in order to grind the vegetation in their stomachs into pulp. The rounded, polished stones called gastroliths were left in a heap where the dinosaur died.

Some dinosaurs might have developed complex mating rituals. Others might have sported elaborate head gear to attract females or used complex vocal resonating devices for mating calls. Dinosaurs appeared to produce sounds by oscillating cartilage in their throats. The larger the cartilage the lower the sound frequency, enabling some species to make a noise somewhat like a foghorn, which carried long distances. One 75-million-year-old herbivore had a 4.5-foot-long hollow crest arching backward from the skull, possibly used as a sexual display device or as a resonating chamber to produce unusual sound effects.

In the Gobi Desert of Mongolia, a spectacular fossil of 9-foot-long, carnivorous dinosaur called an oviraptor (Fig. 137) was discovered in 1993. It was a fleet-footed predator whose name, which literally means "egg hunter," was a misnomer because it was originally thought to have raided nests of other dinosaurs. The fossil oviraptor sat on a nest filled with as many as two dozen eggs neatly laid out in a circle, with the thinner ends pointing to the outside. The dinosaur resembled a wingless version of an ostrich with a shortened neck and a long tail. It sat with its pelvis in the middle of the nest and had its long arms wrapped around the nest the way birds do. It might have been pro-

Figure 137 An oviraptor brooding a clutch of eggs.

tecting the eggs against a gigantic sandstorm that apparently engulfed and fossilized it along with its clutch 70 to 80 million years ago.

Prior to the discovery, nests with eggs and whole infant dinosaurs have been found. However, no direct evidence existed of parents squatting on eggs. The oviraptor was in the exact position a chicken would take sitting on a nest. Whether the dinosaur was keeping the eggs warm as birds do, shading them from the hot sun, or protecting them is unknown. The clutch was probably a communal nest similar to that used by ostriches, into which hens deposit their eggs and take turns incubating them.

The discovery left little doubt that oviraptors brooded because they are so closely related to birds. It is the strongest evidence of parental attention, suggesting birds inherited this behavior from dinosaurs. Oviraptors and birds probably evolved from a common ancestor that also exhibited brooding behavior. The animal apparently lived prior to the first known bird, *Archaeopteryx,* which evolved alongside the dinosaurs, who mysteriously died out while birds live on to this very day.

After a discussion of some of the strangest creatures on Earth, the next chapter will take a look at some of the most unusual places where plants and animals manage to live.

10

LIFE IN THE STRANGEST PLACES

HIDDEN CREATURES OF THE WORLD

This chapter searches for life in some of the most severe living conditions our planet has to offer. The Earth's biosphere, which all living entities call home, is more extensive and contains more abundant life than previously thought possible. Life on the surface is so apparent that humans often forget that most of the world's living species lie hidden well out of sight. Species have adapted to nearly every conceivable environment, from subfreezing to boiling, strongly acidic to toxic alkaline, and have even adapted to extreme pressures of the abyss and far below the ocean floor.

Living bacteria have also been discovered deep underground. Bacterial spores trapped in amber millions of years ago and buried under thick sediments have been brought back to life. On the bottom of the ocean in the cold and dark lies an eerie world occupied by some of the strangest creatures on Earth. The discovery of complex animals, sometimes called extremeophiles, living in such totally unexpected and bizarre habitats shows how resilient life is even under extreme conditions.

LIVING BENEATH THE ICE

Antarctica, called "terra incognita" by early explorers, is indeed a lost continent. All of its geographic features, including mountain ranges, high plateaus, lowland plains, and canyons, are buried under a sheet of ice in places as much as 3 miles thick. Perhaps the most impoverished desert on Earth is surprisingly in the large, ice-free Dry Valleys of Antarctica. Hurricane force winds race unimpeded across the Antarctic plateau, polishing rocks into pyramid-shaped ventifacts and scouring the valleys of all but invisible life-forms.

Only meager signs of life exist, including blue-green algae on the bottoms of small glacier-fed lakes, soil bacteria, and a giant wingless fly. Five inches beneath the valley floor, the dirt teems with tiny nematode worms. In small, ice-covered lakes, mats of algae and plankton live a strange, part-time existence. During the sunless winters, they freeze in a suspended state called cryobiosis. However, when the waters thaw during the short nightless summers, their life reawakens to begin a new cycle.

Antarctica hosts just two flowering plant species. These have undergone population explosions recently, possibly due to a warming climate. Delicate mosses and lichens, if disturbed, take a century to recover. The discovery of lichens in tiny pores on the undersides of rocks has fueled speculation that similar life-forms might inhabit the planet Mars, whose frigid terrain exhibits many similarities to the frozen continent.

By pushing the envelope of life to new extremes, scientists have found evidence that living microbes populate Lake Vostok, a gigantic, near-freezing freshwater lake hidden for hundreds of thousands of years 13,000 feet beneath the Antarctic ice sheet. Drillers retrieved ice cores that represent lake water that froze to the bottom of the glacial sheet. Viable cells of primitive bacteria called archaea were discovered when the ice was melted in the laboratory. Microbes originally living in the lake became trapped in the ice as it froze when Antarctica became glaciated. The organisms probably survived off nutrients spewing out of hot springs on the bottom of the lake.

Around the perimeter of the huge ice continent, sea ice (Fig. 138) expands to 7.7 million square miles in winter, more than twice the size of the United States, growing at an average rate of 22 square miles a minute. During four months of the year, Antarctica is in total darkness. Even during the short summer season, the water under the ice receives less than 1 percent of the sunlight on the surface. The water temperature throughout the year varies from about 1 to 2 degrees Celsius below freezing, but due to a high salt content, it does not ice up. The ice is no more than 3 feet thick, much less than its Arctic counterpart. Yet it is one of the strangest ecosystems on Earth.

The sea surrounding Antarctica is the coldest marine environment and was once though to be totally barren of life. Yet the waters surrounding

Figure 138 *A U.S. Coast Guard icebreaker clearing a path through the ice in McMurdo Sound, Antarctica.*

(Photo by T. F. Ahlgrim, courtesy U.S. Navy)

Antarctica are teeming with life (Fig. 139). The Antarctic Sea represents about 10 percent of the total extent of the world's ocean and is the planet's largest coherent ecosystem. The abundance of species in the polar regions is due, in most part, to their ability to survive in subfreezing water.

During the Antarctic winter, from June to September, sea ice covering much of the ocean around the continent is punctured in places by coastal and ocean polynyas. These are large, open-water areas kept from freezing by upwelling warm-water currents. Antarctic sea ice differs from that in the Arctic, where most of the ocean is surrounded by land, which dampens the seas and allows the ice to grow over twice as thick. Some of the Arctic ice that survives the summer doubles in thickness in four years. In the Antarctic, by comparison, powerful storms at sea churn the water and break up the ice, preventing it from growing any thicker.

As the ice forms, organisms become trapped in it. Instead of hibernating to endure winter's cold, living dynamic organisms grow in astonishing abundance, sometimes in thick, hairy mats on the underside of the ice. Algae known as phytoplankton (Fig. 140) thrive in temperatures 4 degrees below freezing, with less than one-half of 1 percent of the surface light and surrounded by water three times saltier than seawater. They continue growing as nutrient-rich seawater moves into voids in the ice. The influx of food makes the icebound population of algae explode. They stop growing only when the winter sunlight diminishes and the ice becomes colder and freezes entirely.

Figure 139 *A view of marine life found on the bottom of McMurdo Sound, Antarctica.*

(Photo by W. R. Curtsinger, courtesy U.S. Navy)

When longer spring days bring more sunlight, algae, bacteria, and other microscopic organisms began to grow rapidly on the undersides of the sea ice. As the weather warms, the ice begins to break up, and the organisms escape into the water. There they support an escalating food chain of whales,

Figure 140
Phytoplankton such as coccolithospeore maintain living conditions on Earth.

Figure 141 *Tiny organisms like this krill provide food for larger species.*

fish, and seals. The algae show little evidence of damage caused by the annual ozone hole, which forms each September over Antarctica. They appear to weather the onslaught of ultraviolet rays surprisingly well. Perhaps they are protected by the sea ice during the worst of the radiation bombardment.

Several thousand times more organisms live in the ice than in the water just underneath it. Tiny shrimplike crustaceans called krill (Fig. 141) winter beneath the ice, grazing off the ice algae. Krill serve as a major food source for other animals along the food chain on up to whales. Whether they can actually remove algae from under the ice or simply consume the algae melted at the ice-water interface is unknown. The biomass of krill exceeds that of any other animal species, amounting to well over a billion tons.

Many species of herbivorous zooplankton are important links in the marine food chain. Tiny plant-eating marine crustaceans called copepods need openings in the ice to reach the surface, where they can feed on plants and reproduce. However, if the ice is too extensive, few of them will emerge. Reproduction usually begins just before or immediately after the sea ice breaks up, producing a single brood. Infant development occurs rapidly during the melting season, with most individuals reaching adolescence before the onset of new ice.

In the winter months, virtually no phytoplankton inhabit the seawater. Yet in the late winter or early spring, the young copepods miraculously emerge as mature adults. Apparently, the hungry crustaceans winter beneath the ice, grazing off the ice algae. Somehow, they survive the ordeals of the harsh winter and continue to make the Southern Ocean one of the richest environments.

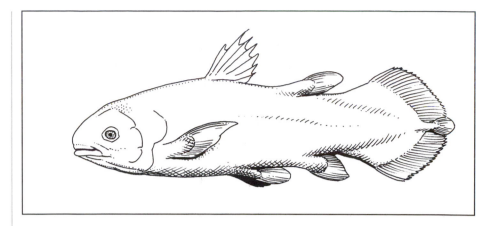

ple, a totally new and unexpected algae lives at depths of some 900 feet, deeper than any previously known marine plant. The species comprises a variety of purple algae with a unique structure. It consists of heavily calcified lateral walls and very thin upper and lower walls to allow for maximum surface exposure to the feeble sunlight at these depths.

In the great depths of the abyss, animals live in the cold and dark, adapting to such high pressures they perish when brought to the surface. Some bacteria along with higher life-forms live successfully at extreme pressures of more than 1,000 atmospheres in the deepest parts of the sea but cannot grow at pressures of less than 300 atmospheres. The bacteria aid in the decay of dead plant and animal materials that fall to the deep seafloor to recycle organic matter in the ocean.

On the sunless floor of the Gulf of Mexico 1,800 feet below the surface waters, marine biologists discovered what appears to be a new, remarkable worm species living among mounds of frozen natural gas, or methane hydrate, that seeped from beneath the seabed. This crystallized blend of water ice and natural gas forms a rocklike mass in great abundance in the high pressures and low temperatures of the deep sea. Estimates indicate that enough methane is locked within hydrates to blanket the entire Earth with a layer of gas 160 feet thick.

The Gulf of Mexico is one of the few places where lumps of hydrate actually break through the seafloor. The worms are flat, pink, centipedelike creatures 1 to 2 inches long. They live in dense colonies that tunnel through the 6-foot-wide ice mounds shaped like mushrooms on the ocean floor. The worms appear to survive by eating bacteria growing on the yellow and white ice mounds. They might also feed directly on the methane within the hydrates. Despite their rock-solid appearance, the hydrates are fairly unstable. Slight changes in temperature or pressure caused by the tunneling worms could cause the methane ice to melt, prompting the seafloor above it to cave in, possibly to the detriment of the worms.

The karst terrain in the jungles of Mexico's Yucatan Peninsula displays a bizarre realm of giant caverns and sinkholes that provide an undersea realm of astonishing beauty. The sinkholes formed when the upper surface of a limestone formation collapsed, exposing the watery world beneath the jungle floor. The caves are linked by miles of twisting passages 100 feet beneath the ocean. Sightless creatures occupying the deepest recesses of the caves are blinded by generations living in utter darkness.

Strange, previously unknown creatures, including spiders, beetles, leeches, scorpions, and centipedes, inhabit the deep, dark passages of Movile Cave 60 feet below ground in southern Romania. The cave is a closed subterranean ecosystem sealed off from the surface and nourished by hydrogen sulfide rising from the Earth's interior. Bacteria at the bottom of the food chain metabolize hydrogen sulfide in a processes called chemosynthesis.

The cave's occupants evolved over the past 5 million years. They live with little oxygen and absolutely no light. As a result, they lack pigmentation and eyesight. The cave, which winds beneath 150 square miles of dry countryside, began when the Black Sea dropped precipitously some 5.5 million years ago. The cave developed in a limestone formation when the waters began rising again. It was sealed off from the outside world when clay impregnated the limestone, making it watertight, and when thick layers of wind-driven sediment were deposited on top during the ice ages.

LIFE IN HOT WATER

Organisms living at high temperatures have long fascinated biologists. Natural hot-water environments are widely distributed throughout the world. They are generally found in association with volcanic activity both on land and under the sea, where sulfur is often plentiful for sulfur-metabolizing bacteria. Primitive bacteria descended from the earliest known form of life and remain by far the most abundant organisms, comprising some 80 percent of the Earth's biomass.

Evidence that life began quite early in Earth history when the planet was steaming hot exists today as archaebacteria, or simply archaea, living in thermal springs and other hot-water environments throughout the world. They range more widely than previously believed. Many parts of the ocean are teeming with them. A third of the biomass of picoplankton (the tiniest plankton) in Antarctic waters are archaea. Such abundance could mean that archaea play an important role in the global ecology and might have a major influence on the chemistry of the ocean.

Archaea seem to flourish only in such inhospitable locales as sulfurous hot springs, isolated pools of brine, and wet, murky sediments lacking oxygen. Half a mile below the desert floor of New Mexico are thick salt deposits evaporated from an ancient sea 250 million years ago. Microbiologists revived bacteria that apparently lived during the late Permian period trapped in the salt crystals. The bacteria can form spores that might allow them to survive for hundreds of millions of years. The microbes derive energy from simple carbon molecules, such as glycerol, acetate, and pyruvate. Other salt-dependent archaea, called halobacteria, reside in such places as the Dead Sea and the Great Salt Lake. Some archaea that tolerate high temperatures also live in strongly acidic environments. The presence of these organisms suggests that bacteria called thermophiles (heat loving) were the common ancestors of all life.

The origin of thermophilic microorganisms has intrigued scientists for more than a century. The conditions on the early Earth would have been ripe for the evolution of thermophilic organisms. Most of these species have a sulfur-based energy metabolism that combines sulfur with hydrogen to form hydrogen sulfide. Sulfur compounds expelled from a profusion of volcanoes would have been abundant on the hot, volcanically active planet. Other thermophilic microorganisms live off organic materials, combining carbon with hydrogen to form methane.

Multicellular plants and animals cannot survive at water temperatures exceeding 50 degrees Celsius. Therefore, only microorganisms are found above these temperatures. The upper temperature limit for microorganisms appears to be 60 or 70 degrees for eukaryotes probably because the nucleus is unable to function at higher temperatures. By comparison, organisms that lack a nucleus, such as archaea and other bacteria, live in most boiling-water environments where they often reproduce extremely well. Apparently, bacteria can grow even in environments above the normal boiling point of water as long as the water remains a liquid. The boiling temperature of water depends on the pressure. Although water boils at 100 degrees at sea level, at the ocean depths it can remain a liquid at several hundred degrees.

Biologic constraints place the upper temperature limit for organisms at about 150 degrees, above which amino acids break down. Therefore, life can conceivably exist at temperature ranges between 100 and 150 degrees. The present record holder is 110 degrees, and evidence of colonies of bacteria living near hydrothermal vents on the ocean floor (Fig. 145) suggests they can grow at 130 degrees or more. Hydrothermal vents located at midocean spreading ridges spew out water at temperatures of several hundred degrees and are thought to be the main route through which the Earth's interior loses heat.

Volcanic chimneys near the Juan de Fuca spreading ridge system in the Pacific yielded microbes that might well be the toughest forms of life ever dis-

Figure 145
Hydrothermal vents on the deep-ocean floor provide nourishment and heat for bottom dwellers.

(Courtesy of USGS)

covered, thriving at temperatures higher than thought possible. The chimneys resemble stalagmites formed when volcanically heated brines spew out of the ocean floor and deposit a variety of sulfide minerals. At the center of these structures, water temperatures can reach more than 350 degrees. On the exterior of these chimneys were a host of heat-loving microorganisms. Some lived quite close to the central conduit at temperatures possibly reaching 200 degrees.

Thermophiles are living fossils from the Earth's earliest days. They were created in a scalding environment about 4 billion years ago, perhaps around volcanic hot springs on the ocean floor. Since the first organisms apparently arose in such high-temperature environments, thermophilic organisms are considered primordial. All subsequent organisms were derived from them. Therefore, the common ancestor of all life on Earth was probably a thermophile.

SUBTERRANEAN LIFE

A study of sediments laid down in the Pacific Ocean more than 4 million years ago revealed bacteria living at depths of 1,700 feet or more beneath the ocean floor. Scientists had thought that temperatures and pressures this far below the seabed would be too extreme for life to establish itself. Yet apparently, the microorganisms grow quite well in the searing heat. The bacteria feed off the organic matter available at such depths and grow much larger than their counterparts living on the seafloor above. They are anaerobic organisms, which would die if exposed to oxygen. Perhaps most surprising is the sheer size of this new biosphere. The mass of subterranean bacteria is possibly equivalent to 10 percent of the total mass of all life on the surface.

Far below ground, locked up in rocks as much as 2 miles down, are microbes that apparently remained prisoners for millions of years. Subsurface bacteria and similar organisms called archaea were originally discovered in oil well cuttings from deep below the surface. The subterranean organisms could have evolved from microbes buried more than 100 million years ago, when the sedimentary rocks encasing them were initially deposited.

The subsurface microbes bear little resemblance to their aboveground relatives. The deep bacteria, known as thermophiles, make their home in rock with an ambient temperature of 75 degrees Celsius. One such bacteria has earned the name *Bacillus infernus* for its hellish habitat. Similar thermophilic microbes colonize volcanically heated springs, such as those of Yellowstone National Park (Fig. 146).

The anaerobic organisms live on a sparse diet of petroleum and other organic compounds buried along with them and barely hold on to life in a

Figure 146 *The largest hot spring in the western group of Snake River Hot Springs, Yellowstone National Park, Wyoming.*

(Photo by J. D. Love, courtesy USGS)

sort of suspended animation. Because the nutrients are so poor, the microbial colonies do not receive sufficient food to grow or reproduce, nor do they have room to spread through the rock. They remain prisoners locked up in the tightly spaced mineral grains unable to enter or leave their underground cages.

In the dry valleys of Antarctica (Fig. 147), microorganisms take shelter from the extreme cold within the outermost layers of rocks. An entirely unexpected community of microscopic algae, fungi, and bacteria was found living inside minute gaps between the grains of sandstone just beneath the surface. They receive little sunlight, moisture, and sustenance from the mineral that encloses them. The tiny fragments of life might be thousands of years old, more ancient than the oldest forests.

Some organisms do inhabit sedimentary rocks, which are often rich in organic compounds. However, other deep communities living within igneous formations, which are extremely hard rocks lacking organic nutrients, are not so fortunate. The fact that granite is one of the most abundant rocks on the continents suggests these organisms are quite widespread. The industrious residents of these rocks fashion their own organic molecules out of the barest of inorganic materials that originate from the Earth's interior. From deep within the Columbia River basalts, scientists have discovered a community of microbes that are apparently sustained by only hydrogen and carbon dioxide carried in the waters of a subsurface aquifer. The microbes process hydrogen, carbon dioxide, and water into the molecules of life, making them totally independent and therefore unique among the vast array of living species.

Life might also be hiding beneath the dusty red surface of Mars (Fig. 148) and bodies even more distant. If primary production can occur underground disconnected from photosynthesis, then subsurface life might exist in other parts of the solar system. The surface of Mars is presently inhospitable because it lacks liquid water. However, nourishing fluids might course through the warmer interior of the planet, allowing life to flourish.

In deep caves bacterial spores have been discovered entombed long ago in fossilized tree sap, or amber. Some microbes might be as much as 135 million years old, having lived during the height of the dinosaur age. In times of crises, many bacteria and fungi can transform themselves into biologic time capsules until conditions of life improve. As spores, the microbes cease moving, eating, and reproducing. They become virtually indestructible, able to survive over lengthy periods without air or water. They can thus withstand the harshest environments.

Other cave creatures consist of bacterial colonies that are apparently unique to a cavern located in the Mexican state of Tabasco about 40 miles south of Villahermosa. A white slime coated the cave walls and ceiling, which

Figure 147 *Wright Dry Valley, Antarctica.*

(Photo by D. R. Thompson, courtesy U.S. Navy)

Figure 148 *The surface of Mars from* Viking 1, *showing boulders surrounded by windblown dust and sand.*

(Photo courtesy NASA)

dripped with water more acidic than battery acid. The slime contained bacteria that ingested sulfur as its energy source and excreted a strong sulfuric acid. The bacteria serve as the bottom of the food chain for an unusual sulfur-based ecosystem inside the mile-long cave.

The weight of the subterranean microbes could equal that of all organisms on the surface above them. They might reach as far as 2.5 miles below the continental crust and more than 4 miles into the oceanic crust. Any deeper and the rock is presumably too hot for life. Their existence also suggests life might even have originated deep down in the bowels of the Earth well protected from the harsh conditions on the surface. In its early days, the planet was pounded by giant meteorites, and the Sun's harmful rays singed the surface. Therefore, any life attempting to evolve above ground would have met a sizzling death.

215

CREATURES OF THE THERMAL VENTS

Beneath the ocean at depths of 8,000 feet lies an eerie world. An oasis of hydrothermal vents and exotic, deep-sea creatures previously unknown to science, including blind shrimps, unusual echinoderms, and metal–eating bacteria, thrive in the cold, dark abyss. The base of jagged basalt cliffs displayed evidence of active lava flows, including fields strewn with pillow lavas as though extruded from a giant toothpaste tube. Perched near the seams of midocean ridges are curious pillars of lava standing like Greek columns 45 feet tall.

Hydrothermal vents called black smokers (Fig. 149) spew hot water blackened with sulfide minerals. The largest known black smoker is a 160-foot-tall structure off the coast of Oregon appropriately named Godzilla after the giant reptile of Japanese science-fiction film fame. Other hydrothermal vents called white smokers eject milky colored hot water. The vents have openings typically ranging from less than a half inch to more than 6 feet across. They commonly occur throughout the world's oceans along the midocean spreading ridge system extending 46,000 miles around the globe like the stitching on a baseball. The undersea geysers built forests of tall chimneys spouting hot water into the near freezing ocean bottom.

Figure 149 A *hydrothermal vent on the East Pacific Rise pouring out black, sulfide-laden hot water.*

(Photo by N. P. Edgar, courtesy USGS)

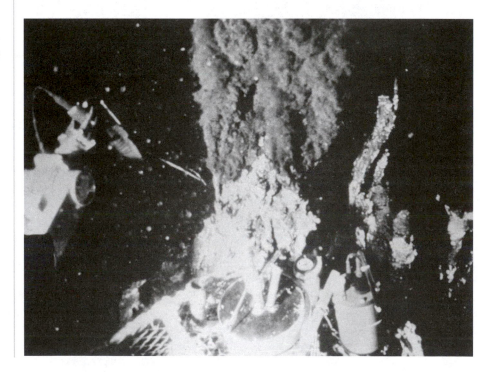

Figure 150 Tall tube worms, giant clams, and large crabs live near hydrothermal vents on the deep-ocean floor.

The hydrothermal vents identified at more than a dozen sites on mid-ocean ridges scattered throughout the world support the world's most bizarre biology. The undersea geysers maintain the bottom waters at tolerable temperatures and provide valuable nutrients to sustain growth. The environment is so unique that it is totally independent of the Sun as a source of energy, which instead is supplied by the Earth itself.

Flourishing among the black smokers are perhaps the strangest creatures ever encountered because of their unusual habitat. Perhaps the first life-forms originated around such vents, obtaining from the Earth's hot interior all the nutrients necessary for survival. In such an environment, life could have evolved a mere half billion years after the creation of the Earth.

Large communities of unusual organisms as crowded as the tropical rain forests cluster around the hydrothermal vents (Fig. 150). Large white clams up to a foot long and mussels, lacking skin pigments, nestle between black pillow lava. Giant white crabs scamper blindly across the volcanic terrain, and long-legged marine spiders roam the ocean floor. The vent creatures live in total darkness and therefore do not need eyes, which for many have become useless appendages. More than 300 new species of vent animals have been identified since the first hydrothermal vent was discovered in 1977.

In the Pacific Ocean, clusters of giant tube worms up to 10 feet tall sway in the hydrothermal currents. The tube worms are contractible animals living inside long, white stalks up to 4 inches wide. They can increase in length at a rate of more than 33 inches per year, making them the fastest-growing marine invertebrates. The worms are so accustomed to the high pressure at these great depths they immediately perish and come out of their tubes when brought to the surface.

While feeding, tube worms expose a long, bright, gill-like, red plume abundantly supplied with blood. The tube worms use these plumes to collect hydrogen sulfide, nitrate, and other nutrients to feed symbiotic bacteria in their guts, which break down these compounds to provide nutrition for the worms. The brown, spongy tissue filling the inside of a tube worm is packed with bacteria, comprising some 300 billion bacteria per ounce of tissue. The plume is also a delicacy for hungry crabs, which attempt to climb the stalks in search of a meal. Tube worms reproduce by spawning, releasing sperm and eggs, which combine in the water to create a new worm complete with its own starter set of bacteria.

Dominating the hydrothermal vents in the Atlantic are swarms of small shrimp. They were originally thought to be blind until the discovery of an unusual pair of vision organs on the shrimps' backs to replace the eyes attached to their heads. Apparently, they see by the feeble light emanating from the hot-water chimneys. Thermal radiation alone cannot explain the light emitted by the 350 degrees C water. The light is possibly produced by the sudden cooling of the hot water. This produces crystalloluminescence as dissolved minerals crystallize and drop out of solution, thereby emitting light. The light, although extremely dim, is apparently bright enough to allow photosynthesis to take place even on the very bottom of the ocean. The deep-sea light has intrigued biologists because of the possibility that organisms can harness this energy by using a type of photosynthesis totally independent of the Sun.

Nearly every food web on the planet is supported by photosynthetic organisms using solar energy to produce carbohydrates. In 1984, during an expedition using the deep-diving submersible *Alvin* (Fig. 151), researchers discovered groups of animals thriving in the extreme depths of the Gulf of Mexico, well beyond the reach of the Sun's rays and far from seafloor hot springs. Communities of mussels, tube worms, crabs, fish, and other marine animals thrived on the cold seafloor at the base of the Florida escarpment. The animals lived in symbiosis with bacteria, which grazed off organic matter disseminated in the porous rock.

Instead of browsing on detrital material falling from above as do most creatures of the deep abyss, the vent animals rely on a symbiotic relationship with sulfur-metabolizing bacteria living within the host's tissues to obtain

Figure 151 *The deep-sea submersible* Alvin *secured to the deck of the USS Fort Snelling.*

(Courtesy U.S. Navy)

nutrition. The bacteria metabolize sulfur compounds in the hydrothermal water by chemosynthesis. They harness energy liberated by the oxidation of hydrogen sulfide supplied by the vents. This enables them to incorporate carbon dioxide for the production of organic compounds such as carbohydrates, proteins, and lipids. The host animal absorbs the by-products of the bacteria's metabolism to nourish itself. The vent animals are so dependent on the bacteria the mussels have only a rudimentary stomach and the tube worms lack even a mouth.

Animals also feed directly on the bacteria alone. Odd-looking colonies of bacteria, some with long tendrils swaying in the warm currents, become the feeding grounds for complex forms of higher life. Drifting bits of whitish bacteria that swirl like falling snow cloud some regions. Clumps of waving bacteria occasionally break loose from fissures and join the swirl of biologic snowfall. But perhaps the most impressive sight of this deep underwater world was a massive blinding snowstorm of bacteria, which formed mats on the ocean floor several inches thick.

The vent creatures live precarious lives because the hydrothermal systems switch on and off sporadically. Therefore, they can survive only as long as the black smokers continue to operate, usually on timescales of perhaps only a few years. To illustrate this point, isolated piles of empty clamshells bear witness to local mass fatalities. In newly formed basalt fields, the vent creatures soon establish residency around young hydrothermal vents. New life quickly colonizes the previously barren abyss, turning it into a rich, undersea garden.

CONCLUSION

Wie humans are relative newcomers on the evolutionary scene. During the last 10,000 years, a period known as the Holocene epoch, synonymous with the rise of civilization, the Earth has been exceedingly beneficial to humankind. Humans have prospered well under a benign environment. Today, however, major changes are taking place. Apparently, our species is responsible for much of the climatic and biological disturbances that beset our planet. Human beings are conducting a global experiment by changing the face of the entire planet. We are rapidly filling the world ourselves. As our species continues to squander the Earth with total disregard for its life-sustaining properties, our actions might force the planet into another mode of operation that will no longer be benevolent to humankind.

This humanmade destruction will eclipse all the geologic hazards faced by humanity since our species first evolved. When considering the great upheavals in the Earth over the last few million years, when great ice sheets spanned the northern continents, and the substantial climate changes that ensued during that time, it is a wonder that our species survived to the present. Many other species that had been around much longer did not endure such disruptions of their environment. If we continue to upset nature's delicate balance through our wanton negligence and waste, we could find ourselves threatened with extinction as well.

GLOSSARY

aa lava (AH-ah)   Hawaiian name for blocky basalt lava

abyss   the deep ocean, generally over a mile in depth

acanthostega (ah-KAN-the-stay-ga)   an extinct primitive Paleozoic amphibian

age   a geologic time interval shorter than an epoch

albedo   the amount of sunlight reflected from an object and dependent on color and texture

alpine glacier   a mountain glacier or a glacier in a mountain valley

amber   fossil tree resin that has achieved a stable state after ground burial due to chemical change and the loss of volatile constituents

ammonite (AM-on-ite)   a Mesozoic cephalopod with flat, spiral shells

amphibian   a cold-blooded, four-footed vertebrate belonging to a class midway in the evolutionary development of fish and reptiles

andesite   a volcanic rock intermediate between basalt and rhyolite

angiosperm (AN-jee-eh-sperm)   flowering plant that reproduces sexually with seeds

annelid (A-nil-ed)   wormlike invertebrate characterized by a segmented body with a distinct head and appendages

archaea (AR-key-ah)   primitive, bacteria-like organism living in high-temperature environments

archaeocyathan (AR-key-ah-sy-a-than)   an ancient Precambrian organism resembling sponges and corals and that built early limestone reefs

Archaeopteryx (AR-key-op-the-riks)   primitive, Jurassic, crow-sized bird characterized by teeth and a bony tail

Archean (AR-key-an)   major eon of the Precambrian from 4.0 to 2.5 billion years ago

arthropod (AR-threh-pod)   the largest group of invertebrates, including crustaceans and insects, characterized by segmented bodies, jointed appendages, and exoskeletons

asteroid   a rocky or metallic body, orbiting the Sun between Mars and Jupiter, and leftover from the formation of the solar system

asteroid belt   a band of asteroids orbiting the Sun between the orbits of Mars and Jupiter

astrobleme   eroded remains on the Earth's surface of an ancient impact structure produced by a large, cosmic body

atmospheric   the weight per unit area of the total mass pressure of air above a given point; also called barometric pressure

Azoic eon   a term applied to the first half billion years of Earth history

Baltica (BAL-tik-ah)   an ancient Paleozoic continent of Europe

barrier island   a low, elongated coastal island that parallels the shoreline and protects the beach from storms

basalt   a dark, volcanic rock rich in iron and magnesium and usually quite fluid in the molten state

basement   the surface beneath which sedimentary rocks are not found; the igneous, metamorphic, granitized, or highly deformed rock underlying sedimentary rocks

batholith   the largest of intrusive igneous bodies, more than 40 square miles on its uppermost surface

belemnite (BEL-em-nite)   an extinct Mesozoic cephalopod with a bullet-shaped internal shell

bicarbonate   an ion created by the action of carbonic acid on surface rocks; marine organisms use the bicarbonate along with calcium to build supporting structures composed of calcium carbonate

biogenic   sediments composed of the remains of plant and animal life such as shells

biomass   the total mass of living organisms within a specific habitat

biosphere   the living portion of the Earth that interacts with all other biological and geologic processes

bivalve   a mollusk with a shell comprising two hinged valves, including oysters, muscles, and clams

black smoker   superheated hydrothermal water rising to the surface at a midocean ridge; the water is supersaturated with metals; when exiting through the seafloor, it quickly cools and the dissolved metals precipitate, resulting in black, smokelike effluent

blastoid   an extinct, Paleozoic echinoderm similar to a crinoid with a body resembling a rosebud

brachiopod (BRAY-key-eh-pod)   marine, shallow-water invertebrate with bivalve shells similar to mollusks and plentiful in the Paleozoic

bryophyte (BRY-eh-fite)   nonflowering plants comprising mosses, liverworts, and hornworts

bryozoan (BRY-eh-zoe-an)   a marine invertebrate that grows in colonies and characterized by a branching or fanlike structure

calcite   a mineral composed of calcium carbonate

calving   formation of icebergs by breaking off of glaciers entering the ocean

Cambrian explosion   a rapid radiation of species that occurred as a result of a large adaptive space, including numerous habitats and a mild climate

carbonaceous (KAR-beh-NAY-shes)   a substance containing carbon, namely sedimentary rocks such as limestone and certain types of meteorites

carbonaceous chondrites   stony meteorites that contain abundant organic compounds

carbonate   a mineral containing calcium carbonate such as limestone and dolostone

carbon cycle   the flow of carbon into the atmosphere and ocean, the conversion to carbonate rock, and the return by volcanoes

Cenozoic (SIN-eh-zoe-ik)   an era of geologic time comprising the last 65 million years

cephalopod (SE-feh-lah-pod)   marine mollusks including squids, cuttlefish, and octopuses that travel by expelling jets of water

chalk   a soft form of limestone composed chiefly of calcite shells of microorganisms

chert   an extremely hard, cryptocrystalline quartz rock resembling flint

chondrule (KON-drule)   rounded granules of olivine and pyroxine found in stony meteorites called chondrites

cladistics   a taxonomic system using evolutionary relationships to classify organisms

class   in systematics, the category of plants and animals below a phylum comprising several orders

climate   the average course of the weather for a certain region over time

coal   a fossil-fuel deposit originating from metamorphosed plant material

coelacanth (SEE-leh-kanth)   a lobe-finned fish originating in the Paleozoic and presently living in deep seas

coelenterate (si-LEN-the-rate)   multicellular marine organisms, including jellyfish and corals

comet   a celestial body believed to originate from a cloud of comets that surrounds the Sun and develops a long tail of gas and dust particles when traveling near the inner solar system

conglomerate a sedimentary rock composed of welded fine-grained and coarse-grained rock fragments

conodont a Paleozoic, toothlike fossil probably from an extinct marine vertebrate

continental glacier an ice sheet covering a portion of a continent

continental drift the concept that the continents drift across the surface of the Earth

continental shelf the offshore area of a continent in shallow sea

continental slope the transition from the continental margin to the deep-sea basin

convection a circular, vertical flow of a fluid medium by heating from below; as materials are heated, they become less dense and rise, cool down, and become more dense and sink

coprolite fossilized excrement, generally black or brown, used to determine the eating habits of animals

coquina (koh-KEY-nah) a limestone comprised mostly of broken pieces of marine fossils

coral a large group of shallow-water, bottom-dwelling marine invertebrates comprising reef-building colonies common in warm waters

correlation the tracing of equivalent rock exposures over distance usually with the aid of fossils

craton the ancient, stable interior region of a continent, usually composed of Precambrian rocks

crinoid (KRY-noid) an echinoderm with a flowerlike body atop a long stalk of calcite disks

crossopterygian (CROS-op-tary-gee-an) extinct Paleozoic fish thought to given rise to terrestrial vertebrates

crust the outer layers of a planet's or a moon's rocks

crustacean (KRES-tay-shen) an arthropod characterized by two pairs of antenna-like appendages forward of the mouth and three pairs behind it, including shrimps, crabs, and lobsters

diapir the buoyant rise of a molten rock through heavier rock

diatom microplants whose fossil shells form siliceous sediments called diatomaceous earth

dinoflagellate (DIE-no-FLA-jeh-late) planktonic, single-celled organisms important in marine food chains

divergent plate the boundary between lithospheric plates where they separate; it generally corresponds to midocean ridges where new crust is formed by the solidification of liquid rock rising from below

dolomite a mineral formed when calcium in limestone is replaced by magnesium

East Pacific Rise   a midocean ridge-spreading system running north–south along the eastern side of the Pacific; the predominant location where hot springs and black smokers were discovered

echinoderm (I-KY-neh-derm)   marine invertebrates, including starfish, sea urchins, and sea cucumbers

echinoid (i-KY-noid)   a group of echinoderms including sea urchins and sand dollars

ecliptic   the plane of the Earth's orbit around the Sun

ecology   the interrelationships between organisms and their environment

ecosphere   the complex interconnections between the biosphere, hydrosphere, atmosphere, and lithosphere

ecosystem   a community of organisms and their environment functioning as complete, self-contained biological unit

Ediacaran   a group of unique, extinct, late Precambrian organisms

environment   the complex physical and biological factors that act on an organism to determine its survival and evolution

eon   the longest unit of geologic time, roughly about a billion years or more in duration

epoch   a geologic time unit shorter than a period and longer than an age

era   a unit of geologic time below an eon, consisting of several periods

erathem   a stratigraphic system consisting of rocks formed during an era

erosion   the wearing away of surface materials by natural agents such as wind and water

esker   a long, narrow ridge of sand and gravel from a glacial outwash stream

eukaryote (yu-KAR-ee-ote)   a highly developed organism with a nucleus that divides genetic material in a systematic manner

eurypterid (yu-RIP-the-rid)   a large, Paleozoic arthropod related to the horseshoe crab

evaporite   the deposition of salt, anhydrite, and gypsum from evaporation in an enclosed basin of stranded seawater

evolution   the tendency of physical and biological factors to change with time

exoskeleton   the hard, outer, protective covering of invertebrates including cuticles and shells

extinction   the loss of large numbers of species over a short duration, sometimes marking the boundaries of geologic periods

extrusive   an igneous volcanic rock ejected onto the Earth's surface

family   in systematics, the category of plants and animals more specific than order and comprising several genera

feldspar   a group of rock-forming minerals comprising about 60 percent of the Earth's crust and an essential component of igneous, metamorphic, and sedimentary rocks

fluvial   stream-deposited sediment

foraminifer (FOR-eh-MI-neh-fer)   a calcium carbonate-secreting organism that lives in the surface waters of the oceans; after death, its shells form the primary constituent of limestone and sediments deposited onto the seafloor

formation   a combination of rock units that can be traced over a distance

fossil   any remains, impressions, or traces in rock of a plant or animal of a previous geologic age

fossil fuel   an energy source derived from ancient plant and animal life that includes coal, oil, and natural gas; when ignited, these fuels release carbon dioxide that was stored in the Earth's crust for millions of years

fulgurite (FUL-je-rite)   a tubular, vitrified crust from fusion of sand by lightning, most common on mountaintops

fumarole   a vent through which steam or other hot gases escape from underground, such as a geyser

fusulinid (FEW-zeh-LIE-nid)   a group of extinct foraminifera resembling a grain of wheat

gastrolith (GAS-tra-lith)   a stone ingested by an animal used to grind food

gastropod (GAS-tra-pod)   a large class of mollusks, including slugs and snails, characterized by a body protected by single shell that is often coiled

genus   in systematics, the category of plants and animals more specific than family and comprising several species

geologic column   the total thickness of geologic units in a region

geothermal   the generation of hot water or steam by hot rocks in the Earth's interior

geyser   a spring that ejects intermittent jets of steam and hot water

glacier   a thick mass of moving ice occurring where winter snowfall exceeds summer melting

glossopteris (GLOS-opt-ter-is)   a late Paleozoic plant that existed on the southern continents but lacking on the northern continents, thereby confirming the existence of Gondwana

gneiss (nise)   a foliated metamorphic rock with similar composition as granite

Gondwana (GONE-wan-ah)   a southern supercontinent of Paleozoic time, comprised of Africa, South America, India, Australia, and Antarctica; it broke up into the present continents during the Mesozoic era

granite   a coarse-grain, silica-rich igneous rock consisting primarily of quartz and feldspars

graptolite (GRAP-the-lite)   extinct Paleozoic planktonic animals resembling tiny stems

graywacke (GRAY-wack-ee)   a coarse, dark gray sandstone

greenstone   a green, weakly metamorphic, igneous rock

greenhouse effect   the trapping of heat in the lower atmosphere principally by water vapor and carbon dioxide

guyot   an undersea volcano that once existed above sea level and whose top was flattened by erosion; later, subsidence caused the volcano to sink below the ocean surface, preserving its flat-top appearance

gypsum   a common, widely distributed mineral frequently associated with halite or rock salt

hallucigenia (HA-loose-ah-gen-ia)   an unusual animal of the early Cambrian with seven pairs of legs and spines along the back

hexacoral   coral with six-sided skeletal walls

hot spot   a volcanic center with no relation to a plate boundary; an anomalous magma generation site in the mantle

hydrocarbon   a molecule consisting of carbon chains with attached hydrogen atoms

hydrologic cycle   the flow of water from the ocean to the land and back to the sea

hydrothermal   relating to the movement of hot water through the crust. It is the circulation of cold seawater downward through the oceanic crust toward the deeper depths of the oceanic crust where it becomes hot and buoyantly rises toward the surface

Iapetus Sea (EYE-ap-i-tus)   a former sea that occupied a similar area as the present Atlantic Ocean prior to the assemblage of Pangaea

ice age   a period of time when large areas of the Earth were covered by massive glaciers

iceberg   a portion of a glacier calved off upon entering the sea

ice cap   a polar cover of ice and snow

ichthyosaur (IK-the-eh-sore)   an extinct Mesozoic aquatic reptile with a streamlined body and long snout

ichthyostega (IK-the-eh-ste-ga)   an extinct primitive, Paleozoic, fishlike amphibian

impact   the point on the surface upon which a celestial object lands

index fossil   a representative fossil that identifies the rock strata in which it is found

interglacial   a warming period between glacial periods

intrusive   a granitic body that invades the Earth's crust

invertebrate   an animal with an external skeleton such as shellfish and insects

iridium   a rare isotope of platinum, relatively abundant on meteorites

island arc   volcanoes landward of a subduction zone, parallel to a trench of a subducting plate and above the melting zone

karst   a terrain comprised of numerous sinkholes in limestone

lacustrine (leh-KES-trene)   inhabiting or produced in lakes

Laurasia (LURE-ay-zha)   a northern supercontinent of Paleozoic time, consisting of North America, Europe, and Asia

Laurentia (LURE-in-tia)   an ancient North American continent

lava   molten magma that flows out onto the surface

limestone   a sedimentary rock composed of calcium carbonate that is secreted from seawater by invertebrates and whose skeletons compose the bulk of deposits

lithosphere   the rocky outer layer of the mantle that includes the terrestrial and oceanic crusts; the lithosphere circulates between the Earth's surface and mantle by convection currents

lithospheric   a segment of the lithosphere, the upper-layer plate of the mantle, involved in the interaction of other plates in tectonic activity

loess   a thick deposit of airborne dust

lungfish   a bony fish that breathes on land and in water

lycopod (LIE-keh-pod)   the first ancient trees of Paleozoic forests; today comprising club mosses and liverworts

lysocline   the ocean depth below which the rate of dissolution just exceeds the rate of deposition of the dead shells of calcareous organisms

lystrosaurus   ancient extinct, mammal-like reptile with large, down-turned tusks

magma   a molten rock material generated within the Earth and that is the constituent of igneous rocks

magnetic field reversal   a reversal of the north-south polarity of the Earth's magnetic poles

mantle   the part of a planet below the crust and above the core, composed of dense rocks that might be in convective flow

marsupial (mar-SUE-pee-al)   a primitive mammal that weans underdeveloped infants in a belly pouch

megaherbivore   a large, plant-eating animal such as an elephant or extinct mastodon

Mesozoic (MEH-zeh-ZOE-ik)   literally the period of middle life, referring to a period between 250 and 65 million years ago

metamorphism   recrystallization of previous igneous, metamorphic, and sedimentary rocks under extreme temperatures and pressures without melting

metazoan   a primitive multicellular animal with cells differentiated for specific functions

meteorite   a metallic or stony celestial body that enters the Earth's atmosphere and impacts onto the surface

methane   a hydrocarbon gas liberated by decomposing organic matter and a major constituent of natural gas

microfossil   a fossil that must be studied with a microscope; used for dating drill cuttings

Mid-Atlantic Ridge   the seafloor-spreading ridge that marks the extensional edge of the North and South American plates to the west and the Eurasian and African plates to the east

midocean ridge   a submarine ridge along a divergent plate boundary where a new ocean floor is created by the upwelling of mantle material

mold   an impression of a fossil shell or other organic structure made in encasing material

mollusk (MAH-lusc)   a large group of invertebrates, including snails, clams, squids, and extinct ammonites, characterized by an internal and external shell surrounding the body

monotreme   egg-laying mammals including platypus and echidna

moraine   a ridge of erosional debris deposited by the melting margin of a glacier

nautiloid (NOT-eh-loid)   shell-bearing cephalopods abundant in the Paleozoic, with only the nautilus surviving

Neogene   the Miocene and Pliocene epochs of the Cenozoic

nutrient   a food substance that nourishes living organisms

oolite (OH-eh-lite)   small, rounded grains in limestone

ophiolite (OH-fi-ah-lite)   oceanic crust thrust upon continents by plate tectonics

orogeny (oh-RAH-ja-nee)   an episode of mountain building by tectonic activity

ozone   a molecule consisting of three atoms of oxygen in the upper atmosphere and that filters out harmful ultraviolet radiation from the Sun

Paleogene   the Paleocene, Eocene, and Oligocene epochs of the Cenozoic

paleomagnetism   the study of the Earth's magnetic field, including the position and polarity of the poles in the past

paleontology (PAY-lee-ON-tah-logy)   the study of ancient life-forms, based on the fossil record of plants and animals

Paleozoic (PAY-lee-eh-ZOE-ic)   the period of ancient life, between 570 and 250 million years ago

Pangaea (PAN-gee-ah)   a Paleozoic supercontinent that included all the lands of the Earth

Panthalassa (PAN-the-lass-ah)   the global ocean that surrounded Pangaea

period   a division of geologic time longer than an epoch and included in an era

permafrost   permanently frozen ground in the Arctic regions

photosynthesis   the process by which plants form carbohydrates from carbon dioxide, water, and sunlight

pH scale   a logarithmic scale depicting the acidity or alkalinity of a substance

phyla   groups of organisms that share similar body forms

phytoplankton   marine or freshwater microscopic, single-celled, freely drifting plant life

placoderm   an extinct class of chordates, fish with armorlike plates and articulated jaws

plate tectonics   the theory that accounts for the major features of the Earth's surface in terms of the interaction of lithospheric plates

prebiotic   conditions on the early Earth prior to the introduction of life processes

precipitation   the deposition of minerals from seawater

primary producer   the lowest member of a food chain

primordial   pertaining to the primitive conditions that existed during early stages of development

prokaryote (pro-KAR-ee-ote)   a primitive organism that lacks a nucleus

protist (PRO-tist)   a unicellular organism, including bacteria, protozoans, algae, and fungi

pseudofossil   a fossil-like body such as a concretion

pterosaur (TER-eh-sore)   an extinct, Mesozoic flying reptile with batlike wings

radiolarian   a microorganism with shells made of silica comprising a large component of siliceous sediments

radiometric dating   determining the age of an object by radiometrically and chemically analyzing its stable and unstable radioactive elements

redbed   a sedimentary rock cemented with iron oxide

reef   the biological community that lives at the edge of an island or continent; the shells from dead organisms form a limestone deposit

regression   a fall in sea level, exposing continental shelves to erosion

reptile   an air-breathing, cold-blooded animal usually covered with scales that lays eggs on dry land

Rodinia   a Precambrian supercontinent whose breakup sparked the Cambrian explosion of species

sandstone   a sedimentary rock consisting of sand grains cemented together

schist (shist)   a finely layered metamorphic, crystalline rock easily split along parallel bands

seafloor spreading   a theory that the ocean floor is created by the separation of lithospheric plates along midocean ridges, with new oceanic crust formed from mantle material that rises to fill the rift

seamount   a submarine volcano that never reaches the surface of the sea

shale   a fine-grained, fissile, sedimentary rock of consolidated mud or clay

shield   areas of the exposed Precambrian nucleus of a continent

species   groups of organisms that share similar characteristics and are able to breed among themselves

spherules   small, spherical, glassy grains found on certain types of meteorites, on lunar soils, and at large meteorite impact sites

strata   layered rock formations; also called beds

stromatolite (STRO-mat-eh-lite)   a calcareous structure built by successive layers of bacteria or algae and that has existed for the past 3.5 billion years

subduction zone   a region where an oceanic plate dives below a continental plate into the mantle; ocean trenches are the surface expression of a subduction zone

tectonic activity   the formation of the Earth's crust by large-scale movements throughout geologic time

tektites   small, glassy minerals created from the melting of surface rocks by the impact of a large meteorite

tephra (TE-fra)   solid material ejected into the air by a volcanic eruption

terrestrial   all phenomena pertaining to the Earth

Tethys Sea (THE-this)   the hypothetical, midlatitude region of the oceans separating the northern and southern continents of Laurasia and Gondwana several hundred million years ago

tetrapod   a four-footed vertebrate

thecodont (THEE-keh-daunt)   an ancient, primitive reptile that gave rise to dinosaurs, crocodiles, and birds

therapsid (the-RAP-sid)   an ancient reptile ancestor of the mammals

therian   animals that have live births such as mammals

thermophilic   relating to primitive organisms that live in hot-water environments

tide   a bulge in the ocean produced by the Sun's and Moon's gravitational forces on the Earth's oceans; the rotation of the Earth beneath this bulge causes the rising and lowering of the sea level

till   nonstratified material deposited directly by glacial ice as it recedes and is consolidated into tillite

transgression   a rise in sea level that causes flooding of the shallow edges of continental margins

trilobite (TRY-leh-bite)   an extinct, marine arthropod characterized by a body divided into three lobes, each bearing a pair of jointed appendages, and a chitinous exoskeleton

tufa   incrustation of calcium carbonate around a spring or vent

tuff   a consolidated, fine-grained, pyroclastic volcanic rock

tundra   permanently frozen ground at high latitudes

type section   a sequence of strata that was originally described as constituting a stratigraphic unit and that serves as a standard of comparison for identifying similar, widely separated units

ultraviolet   the invisible light with a wavelength shorter than visible light and longer than X rays

uniformitarianism   a theory that the slow processes that shape the Earth's surface have acted essentially unchanged throughout geologic time

upwelling   the upward convection of water currents

varves   thinly laminated lake bed sediments deposited by glacial meltwater

vertebrates   animals with an internal skeleton, including fish, amphibians, reptiles, and mammals

BIBLIOGRAPHY

HISTORICAL GEOLOGY

Allegre, Claude J. and Stephen H. Schneider. "The Evolution of the Earth." *Scientific American* 271 (October 1994): 66–75.

de Duve, Christian. "The Birth of Complex Cells." *Scientific American* 274 (April 1996): 50–57.

Horgan, John. "In The Beginning." *Scientific American* 264 (February 1991): 117–125.

Knoll, Andrew H. "End of the Proterozoic Eon." *Scientific American* 265 (October 1991): 64–73.

Levinton, Jeffrey S. "The Big Bang of Animal Evolution." *Scientific American* 267 (November 1992): 84–91.

Monastersky, Richard. "The Rise of Life on Earth." *National Geographic* 194 (March 1998): 54–81.

O'Hanlon, Larry. "Age of the Giant Insects." *Earth* 4 (October 1995): 10.

Orgel, Leslie E. "The Origin of Life on the Earth." *Scientific American* 271 (October 1994): 77–83.

Radetsky, Peter. "Life's Crucible." *Earth* 7 (February 1998): 34–41.

Simpson, Sarah. "Life's First Scalding Steps." *Science News* 155 (January 9, 1999): 24–26.

Tattersall, Ian. "Once We Were Not Alone." *Scientific American* 282 (January 2000): 56–62.

Vogel, Shawna. "Living Planet." *Earth* 5 (April 1996): 27–35.

Waldrop, Mitchell M. "Goodbye to the Warm Little Pond?" *Science* 250 (November 23, 1990): 1078–1080.

Wess, Peter. "Land Before Time." *Earth* 8 (February 1998): 29–33.

York, Derek. "The Earliest History of the Earth." *Scientific American* 268 (January 1993): 90–96.

SEA LIFE

Forey, Peter and Philippe Janvier. "Agnathans and the Origin of Jawed Vertebrates." *Nature* 361 (January 14, 1993): 129–133.

Irion, Robert. "Parsing the Trilobites' Rise and Fall." *Science* 280 (June 19, 1998): 1837.

Kerr, Richard A. "Evolution's Big Bang Gets Even More Explosive." *Science* 261 (September 3, 1993): 1274–1275.

Knauth, Paul. "Ancient Sea Water." *Nature* 362 (March 25, 1993): 290–291.

Landman, Neil H. "Luck of the Draw." *Natural History* 100 (December 1991): 68–71.

Monastersky, Richard. "The Whale's Tale." *Science News* 156 (November 6, 1999): 296–298.

————. "Jump-Start for the Vertebrates." *Science News* 149 (February 3, 1996): 74–75.

Novacek, Michael J. "Whales Leave the Beach." *Nature* 368 (April 28, 1994): 807.

Schreeve, James. "Are Algae—Not Coral—Reefs' Master Builders?" *Science* 271 (February 2, 1996): 597–598.

Svitil, Kathy A. "It's Alive, and It's a Graptolite." *Discover* 14 (July 1993): 18–19.

Vermeij, Geerat J. "The Biological History of a Seaway." *Science* 260 (June 11, 1993): 1603–1604.

Zimmer, Carl. "Breathe Before You Bite." *Discover* 17 (March 1996): 34.

LAND LIFE

Burgin, Toni, et al. "The Fossils of Monte San Giorgio." *Scientific American* 260 (June 1989): 74–81.

Fischman, Josh. "Dino Hunter." *Discover* 20 (May 1999): 72–78.

Grimaldi, David A. "Captured in Amber." *Scientific American* 274 (April 1996): 84–91.

Padian, Kevin and Luis M. Chiappe. "The Origin of Birds and Their Flight." *Scientific American* 218 (February 1998): 38–47.

Pendick, Daniel. "The Mammal Mother Lode." *Earth* 4 (April 1995): 20–23.

Robbins, Jim. "The Real Jurassic Park." *Discover* 12 (March 1991): 52–59.

Schmidt, Karen. "Rise of the Mammals." *Earth* 5 (October 1996): 20–21 & 68–69.

Schueller, Gretel. "Mammal 'Missing Link' Found." *Earth* 7 (April 1998): 9.

Simpson, Sarah. "Wrong Place, Wrong Time, Right Mammal." *Earth* 7 (April 1998): 22.

Storch, Gerhard. "The Mammals of Island Europe." *Scientific American* 266 (February 1992): 64–69.

Vickers-Rich, Patricia and Thomas Hewitt Rich. "Polar Dinosaurs of Australia." *Scientific American* 269 (July 1993): 49–55.

Waters, Tom. "Greetings from Pangaea." *Discover* 13 (February 1992): 38–43.

Wellnhofer, Peter. "Archaeopteryx." *Scientific American* 262 (May 1990): 70–77.

Zimmer, Carl. "Coming onto the Land." *Discover* 16 (June 1995): 120–127.

MASS EXTINCTIONS

Benton, Michael J. "Late Triassic Extinctions and the Origin of the Dinosaurs." *Science* 260 (May 7, 1993): 769–770.

Erwin, Douglas H. "The Mother of Mass Extinctions." *Scientific American* 275 (July 1996): 72–78.

Fields, Scott. "Dead Again." *Earth* 4 (April 95): 16.

Flannery, Tim. "Debating Extinction." *Science* 283 (January 8, 1999): 182–183.

Kerr, Richard A. "Greatest Extinction Looks Catastrophic." *Science* 280 (May 15, 1998): 1007.

———. "The Earliest Mass Extinction?" *Science* 257 (July 31, 1992): 612.

Monastersky, Richard. "Sudden Death Decimated Ancient Oceans." *Science News* 146 (July 16, 1994): 38.

Pendick, Daniel. "The Greatest Catastrophe." *Earth* 6 (February 1997): 34–35.

Raup, David. "Biological Extinction in Earth History." *Science* 231 (March 28, 1986): 1528–1533.

Sullivant, Rosemary. "Flapping Through the Bottleneck." *Earth* 6 (December 1997): 22–23.

Waters, Tom. "Death by Seltzer." *Discover* 18 (January 1997): 54–55.

CAUSES OF EXTINCTION

Alvarez, Walter and Frank Asaro. "An Extraterrestrial Impact." *Scientific American* 263 (October 1990): 78–84.

Courtillot, Vincent E. "A Volcanic Eruption." *Scientific American* 263 (October 1990): 85–92.

Desonie, Dana. "The Threat from Space." *Earth* 5 (August 1996): 25–31.

Gehrels, Tom. "Collisions with Comets and Asteroids." *Scientific American* 274 (March 1996): 54–59.

Gould, Stephen J. "An Asteroid to Die For." *Discover* 10 (October 1989): 60–65.

Grieve, Richard A. F. "Impact Cratering on the Earth." *Scientific American* 262 (April 1990): 66–73.

Hildebrand, Alan R. and William V. Boynton. "Cretaceous Ground Zero." *Natural History* 104 (June 1991): 47–52.

Hoffman, Paul F. and David P. Schrag. "Snowball Earth." *Scientific American* 282 (January 2000): 68–75.

Kerr, Richard A. "Did an Ancient Deep Freeze Nearly Doom Life?" *Science* 281 (August 28, 1998): 1259–1261.

———. "A Volcanic Crisis for Ancient Life?" *Science* 270 (October 6, 1995): 27–28.

Monastersky, Richard. "Eruptions Cleared Path for Dinosaurs." *Science News* 155 (April 24, 1999): 260.

———. "Cretaceous Die-Offs: A Tale of Two Comets?" *Science News* 143 (April 3, 1993): 212–213.

Morell, Virginia. "How Lethal Was the K-T Impact?" *Science* 261 (September 17, 1993): 1518–1519.

Pearce, Fred. "Ice Ages: The Peat Bog Connection." *New Scientist* 144 (December 3, 1994): 18.

Stone, Richard. "The Last Great Impact on Earth." *Discover* 17 (September 1996): 60–71.

Waters, Tom. "Death by Seltzer." *Discover* (January 1997): 54–55.

EFFECTS OF EXTINCTION

Benton, Michael J. "Interpretations of Mass Extinction." *Nature* 314 (April 11, 1985): 496–497.

Ehrlich, Paul R. and Edward O. Wilson. "Biodiversity Studies: Science and Policy." *Science* 253 (August 16, 1991): 758–761.

Eldredge, Niles. "What Drives Evolution?" *Earth* 5 (December 1996): 34–37.

Jablonski, David. "Background and Mass Extinctions: The Alteration of Macroevolutionary Regimes." *Science* 231 (January 10, 1986): 129–133.

Kerr, Richard A. "Origins and Extinctions: Paleontology in Chicago." *Science* 257 (July 14, 1992): 486–487.

Levine, Norman D. "Evolution and Extinction." *BioScience* 39 (January 1989): 38.

Lewin, Roger. "Mass Extinctions Select Different Victims." 231 (January 17, 1986): 219–220.

———. "Extinctions and the History of Life." *Science* 221 (September 2, 1983): 935–937.

May, Robert M. "How Many Species Inhabit the Earth?" *Scientific American* 267 (October 1992): 42–48.

Monastersky, Richard. "Eruptions Spark Explosions of Life." *Science News* 148 (July 1, 1995): 4.

Myers, Norman. "Mass Extinction and Evolution." *Science* 278 (October 24, 1997): 597–598.

———. "Extinction Rates Past and Present." *BioScience* 39 (January 1989): 39–40.

Weissman, Paul R. "Are Periodic Bombardments Real?" *Sky & Telescope* 79 (March 1990): 266–270.

EVOLUTION OF SPECIES

Beardsley, Tim. "Punctuated Equilibrium." *Scientific American* 262 (March 1990): 36–38.

Cartmill, Matt. "Oppressed by Evolution." *Discover* 19 (March 1998): 78–83.

Culotta, Elizabeth. "Ninety Ways to Be a Mammal." *Science* 266 (November 18, 1994): 1161.

Flannery, Tim. "The Case of the Missing Meat Eaters." *Natural History* 102 (June 1993): 41–44.

Gould, Stephen Jay. "The Evolution of Life on the Earth." *Scientific American* 271 (October 1994): 85–91.

Monastersky, Richard. "Eruptions Spark Explosions of Life." *Science News* 148 (July 1, 1995): 4.

Morell, Virginia. "Sizing Up Evolutionary Radiations." *Science* 274 (November 29, 1996): 1462–1463.

Moxon, E. Richard and Christopher Wills. "DNA Microsatellites: Agents of Evolution?" *Scientific American* 280 (January 1999): 94–99.

Peterson, Ivars. "Modeling the Varied Avalanches of Evolution." *Science News* 145 (March 26, 1994): 197.

Shell, Ellen R. "Waves of Creation." *Discover* 14 (May 1993): 54–61.

Waldrop, Mitchell M. "Spontaneous Order, Evolution and Life." *Science* 247 (March 30, 1990): 1543–1545.

Watson, Andrew. "Will Fossil From Down Under Upend Mammal Evolution?" *Science* 278 (November 21, 1997): 1401.

THE LIFE CYCLES

Barinaga, Marcia. "New Clues Found to Circadian Clocks—Including Mammals." *Science* 276 (May 16, 1997): 1030–1031.

Berner, Robert A. and Antonio C. Lasaga. "Modeling the Geochemical Carbon Cycle." *Scientific American* 260 (March 1989): 74–81.

Broecker, Wallace S. and George H. Denton. "What Drives Glacial Cycles." *Scientific American* 262 (January 1990): 49–56.

Green, D. H., S. M. Eggins, and G. Yaxley. "The Other Carbon Cycle." *Nature* 365 (September 16, 1993): 210–211.

Kerr, Richard A. "No Longer Willful, Gaia Becomes Respectable." *Science* 240 (April 22, 1988): 393–395.

———. "The Moon Influences Western U.S. Drought." *Science* 224 (May 11, 1984): 587.

Kunzig, Robert. "Ice Cycles." *Discover* 10 (May 1989): 74–79.

Lo Presto, Charles. "Looking Inside the Sun." *Astronomy* 17 (March 1989): 22–30.

Monastersky, Richard. "New Beat Detected in the Ice Age Rhythm." *Science News* 147 (February 25, 1995): 118.

———. "Ancient Tidal Fossils Unlock Lunar Secrets." *Science News* 146 (September 10, 1994): 165.

Nance, R. Damian, Thomas R. Worsley, and Judith B. Moody. "The Supercontinent Cycle." *Scientific American* 259 (July 1988): 72–79.

Williams, George E. "The Solar Cycle in Precambrian Time." *Scientific American* 255 (August 1986): 88–96.

Zimmer, Carl. "The Case of the Missing Carbon." *Discover* 1 (December 1993): 38–39.

UNIQUE LIFE-FORMS

Beardsley, Tim. "Weird Wonders." *Scientific American* 266 (June 1992): 30–34.

Fischman, Joshua. "Were Dinos Cold-Blooded After All? The Nose Knows." *Science* 270 (November 3, 1995): 735–736.

Gibbons, Ann. "Dino Embryo Recasts Parents' Image." *Science* 266 (November 4, 1994): 731.

Hadfield, Peter. "*T. rex* Growled Like a Gippy Tummy." *New Scientist* 139 (September 11, 1993): 7.

Miller, Mary K. "Tracking Pterosaurs." *Earth* 6 (October 1997): 20–21.

Monastersky, Richard. "Life Grows Up." *National Geographic* 194 (April 1998): 100–115.

———. "Nesting Dinosaur Discovered in Mongolia." *Science News* 149 (January 6, 1996): 7.

———. "The Edicaran Enigma." *Science News* 148 (July 9, 1995): 28–30.

Morell, Virginia. "Warm-Blooded Dino Debate Blows Hot and Cold." *Science* 265 (July 8, 1994): 188.

Morris, S. Conway. "Burgess Shale Faunas and the Cambrian Explosion." *Science* 246 (October 20, 1989): 339–345.

Novacek, Michael J., et al. "Fossils of the Flaming Cliffs." *Scientific American* 271 (December 1994): 60–69.

Stolzenburg, William. "When Life Got Hard." *Science News* 138 (August 25, 1990): 120–123.

Wright, Karen. "When Life Was Odd." *Discover* 18 (March 1997): 52–61.

LIFE IN THE STRANGEST PLACES

Barinaga, Marcia. "Archaea and Eukaryotes Grow Closer." *Science* 264 (May 27, 1994): 1251.

Cann, Joe and Cherry Walker. "Breaking New Ground on the Ocean Floor." *New Scientist* 139 (October 30, 1993): 24–29.

Fischman, Josh. "In Search of the Elusive Megaplume." *Discover* 20 (March 1999): 108–115.

Franklin, Carl. "'Black Smokers' Multiply on Ocean Floor." *New Scientist* 144 (October 22, 1994): 20.

Krajick, Kevin. "To Hell and Back." *Discover* 20 (July 1999): 76–82.

Kunzig, Robert. "Invisible Garden." *Discover* 11 (April 1990): 67–74.

Monastersky, Richard. "Deep Dwellers: Microbes Thrive Far Below Ground." *Science News* 151 (March 29, 1997): 192–193.

———. "Light at the Bottom of the Ocean." *Science News* 145 (January 1, 1994): 14.

Orange, Daniel L. "Mysteries of the Deep." *Earth* 4 (December 1996): 42–45.

Parfit, Michael. "Timeless Valleys of the Antarctic Desert." *National Geographic* 194 (October 1998): 120–135.

Pool, Robert. "Pushing the Envelope of Life." *Science* 247 (January 12, 1990): 158–160.

Stevens, Jane E. "Life on a Melting Continent." *Discover* 16 (August 1995): 71–75.

Stover, Dawn. "Creatures of the Thermal Vents." *Popular Science* 246 (May 1995): 55–57.

INDEX

Boldface page numbers indicate extensive treatment of a topic. *Italic* page numbers indicate illustrations or captions. Page numbers followed by *m* indicate maps; *t* indicate tables; *g* indicate glossary.